Physical Setting Earth Science Laboratory Investigation Handbook

Acknowledgements

I'd like to thank the following for without whom this book would not have been. Thanks to Eric for his patience and help. I'd like to also thank Leah and Tim for their motivation, interest in science and tech support. I'd like to also thank those usual Earthies especially Doc, Gin and Maggie for ideas and inspiration on dry land and otherwise. I'd like to also give a nod to all of those champions of science education that we know from meetings, conferences and cyberspace who might not be listed but did inspire creativity in sharing.

The author would like to thank the following people for their assistance:
Ken Abbott, Renée Aubry, Bill Docekal, Julieen Kane and Keith Williams.

Cover Photo
Depicted is a Cretaceous Seas Diorama found in the Milstein Hall of Ocean Life
AMERICAN MUSEUM OF NATURAL HISTORY
79th @ Central Park West
New York, NY 10024

P. O. Box 328
Onsted, MI 49265-0328
www.topicalrbc.com

ABOUT THIS LABORATORY HANDBOOK

Welcome to Earth Science. You are about to embark on an exciting course of study to help you understand the workings of your home planet. Earth Science is an applied science. That means you will learn about your home using many different science disciplines. There will be physical science, geography, chemistry, and physics.

There are forty investigations in this book related to and intrinsically linked to the Physical Setting Earth Science. Each investigation is just that, a puzzle that must be solved with information from your knowledge of Earth Science.

Each investigation is followed by ten questions. Each question is more in depth than the previous one. All questions are to be answered in complete sentences. Single word answers do not fully explain your understanding of the material.

Terms to know for the Handbook:

Essential Question: What is this investigation about? Why are we conducting the experiment? An example of the essential question is "Why is the sky blue?"

Hypothesis: A Hypothesis is a suggested relationship of two variables. The hypothesis attempts to answer the essential question. An excellent format is the "If…then…because".

> *"If the various qualities of the minerals are tested then the minerals can be identified because minerals have very specific qualities."*

Variables: There are two types of variables which are tested. The *independent variable* is the variable that is being manipulated. The *dependent variable* is the variable that changes as the result of the change in the independent variable. If you were conducting an experiment in the Living Environment on plant growth, the *amount* of water would be the *independent variable* because you control it. The plant *growth* would be the *dependent variable*. Time is a quality of the independent variable.

Free Response Questions: The free response questions are based on the results of your investigation as well as the pre-lab readings. Some questions call on your knowledge of Earth Science and your ability to read the Earth Science Reference Table.

The author:
Rosemarie Sanders,
Earth Science Teacher
Subject Area Representative, Earth Science for
Science Teachers Association of New York State, Westchester Section

Physical Setting Earth Science Laboratory Investigation Handbook

Table of Contents

Physical Setting Earth Science Laboratory Investigation Handbook

Table of Contents

NAME ____________________

1 Bicycle Race

Peter and Steve set out to have a race. Peter is an excellent cyclist and accepts Steve's challenge. Both cyclists know that they do not have a steady rate of riding so they tried to figure out how to determine their patterns.

Peter and Steve recorded, in a chart, their interval time for each kilometer of the race. Looking at the chart, it was hard to determine the rate of change. Steve suggested that they graph their results. Steve figured that if the times were recorded, then the distance between the cyclists would vary because of the different rates the cyclists traveled.

Peter set up the graph with distance along the x-axis and time along the y-axis. Peter explained he would see the travel time better if the graph were constructed this way. The distance scale is in two-tenths of a kilometer and the time scale is in twenty second intervals. He then plotted the data points for each cyclist, connecting the dots as he went using a straight edge.

Peter was struggling with a rate of change calculation. Steve reminded Peter that when working with time as a variable, the seconds, minutes, and hours must be calculated carefully; there are 60 seconds in a minute, and 60 minutes in an hour. It is easier to convert the minutes to seconds by multiplying the minutes by 60 then adding the seconds. It can then be converted back after the equation was done.

Objective: The data will be plotted on a graph so that the rate of change can be visualized.

Essential Question:

____

____

Hypothesis:

If____________________

Then____________________

Because____________________

Variables:

Independent:____________________

Dependent:____________________

Materials: • Data chart • Graph on next page

Directions:

Peter and Steve embark on a bicycle race. The race is a 10 km race. Plot on the graph below the information from the data table. The times listed are as minutes and seconds.

Km	1	2	3	4	5	6	7	8	9	10
Peter	2:20	4:00	5:40	7:00	8:20	9:20	10:20	11:20	12:10	12:50
Steve	4:00	7:20	10:00	12:40	15:00	17:00	19:00	20:40	22:20	23:20

Title: ______________________________

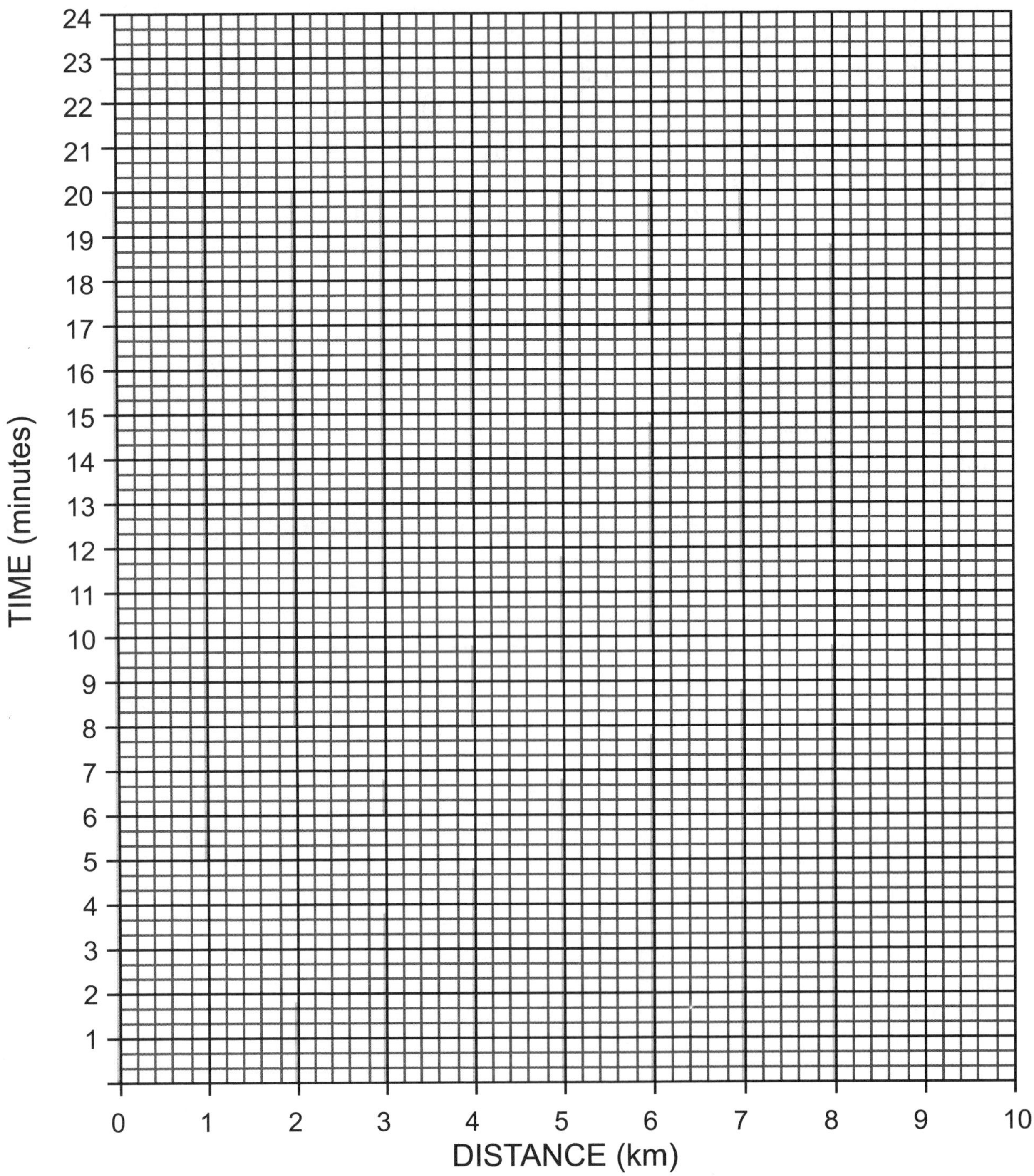

1. What was Peter's rate of change for the bicycle race?

____

____

2. What was Steve's rate of change?

____

____

3. What was the rate of change for Peter between 4 minutes and 7 minutes?

____

4. What was the rate of change for Steve between 4 minutes and 7 minutes?

____

5. Who cycled at a faster rate? Explain how the graph shows the rate of change.

____

____

6. Draw a line ***on the graph*** along the line that indicates 5 minutes, twenty seconds.

   How far did Peter go? ____________ km. How far did Steve go? ____________km

   What was the distance between them?______________________________

7. Draw a line ***on the graph*** along the line that indicates the 5.4 km.

   Peter's time _________ min _________ sec Steve's time _________ min _________ sec

   How much time had to elapse between the two cyclists? ____________________

8. If the time was 12:18 pm when Peter passes the 1.4 km mark, when did the race begin?

   ____

9. Steve claimed he had a slow start and could not keep the pace. Use the data from the rate of change to argue or defined this.

____

____

____

____

10. Explain how a graph can show a rate of change more easily that a data chart.

____

____

____

____

NAME ________________________

2 Density of Liquids

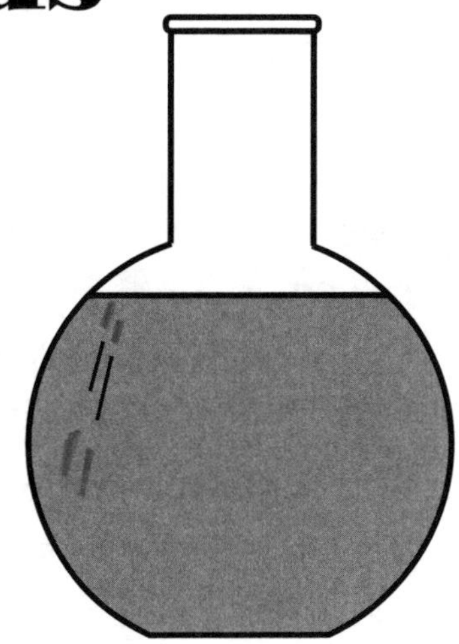

Density is how much matter an object has in a specific space or the ratio of the mass of a substance to its volume. Density can be calculated by dividing the mass by the volume. It is written in units of grams per cubic centimeter (g/cm^3) or grams per milliliter (g/mL). Density also describes the ability of a substance to sink in water. Dense materials like lead and rock sink.

Water has the density of 1 gram per cubic centimeter at 4°C. Specific gravity is the ratio of the density of an object as related to the density of water. The ratio is (the density of a given object):1. Since this density is more or less constant, we can say that water is ~1g/cm^3. If an object has a density of less than 1 g/cm^3, like wood, it will float. If the density is greater than 1 g/cm^3, like most rocks, it will sink. In this lab, you will be finding the density of various liquids. You also will be making *inferences* from *observations* about the graduated cylinder.

Objective:
The objective of this laboratory exercise is to determine the density of liquids.

Essential Question:

____

____

Hypothesis:

If ____

Then ____

Because ____

Variables:

Independent: ____

Dependent: ____

Materials:

- 2 - 50 mL graduated cylinders
- Balance
- 40 mL water
- 40 mL vegetable oil
- 100 mL beaker
- Ice cubes

Procedure: Follow the directions and fill in the chart at the bottom of the page. Show all of your work for calculating density in the space provided. Round answers to the nearest tenth decimal place.

Water

1. Find the mass of an empty graduated cylinder using a balance. Record the answer in column 2.
2. Use a graduated cylinder and carefully measure 40 mL of water into it. Place it on the balance and record the mass in column 1.
3. Calculate the mass of the water by subtracting column 2 from column 1. Record your answer in column 3.
4. Calculate the density of the water. Record your answer in column 5.

$$\text{Density} = \frac{\text{mass}}{\text{volume}}$$

Oil

1. Find the mass of an empty graduated cylinder using a balance. Record the answer in column 2.
2. Use a graduated cylinder and carefully measure 40 mL of oil into it. Place it on the balance and record the mass in column 1.
3. Calculate the mass of the oil by subtracting column 2 from column 1. Record your answer in column 3.
4. Calculate the density of the oil. Record your answer in column 5.

$$\text{Density} = \frac{\text{mass}}{\text{Volume}}$$

	Column 1	**Column 2**	**Column 3**	**Column 4**	**Column 5**
Liquid in graduated cylinder	**Mass of graduated cylinder with Liquid (g)**	**Mass of Empty graduated cylinder (g)**	**Mass of Liquid (g)**	**Volume of Liquid (mL)**	**Density of Liquid (g/mL)**
Water				**40 mL**	
Oil				**40 mL**	

Observations:

1. Which has a greater density, oil or water?

2. Infer: What would happen if you placed an ice cube into the beaker with water? The beaker with oil?

3. Place ice into beakers and describe what you see.

4. Pour all liquids into the 100 ml beaker. Stir. Describe what happens.

5. As the volume of an object increases, without change in temperature or pressure, the density (circle one)

   increases **stays the same** **decreases**

6. As the mass of an object increases, without change in temperature or pressure, the density (circle one)

   increases **stays the same** **decreases**

7. If ice floats in liquid water, what can we infer about the density of the ice?

8. What are the units for density in the metric system? Why?

9. According to the Earth Science Reference Tables in the back of the book, what is the temperature of water when it's density is 1 g/cm^3?

Conclusion:

10. In completing this lab, what have you learned about specific gravity and density?

NAME ______________________________

3 Isolines

The mountains, hills, valleys, gorges and cliffs on a topographic map are known as relief. The contour lines stand for points in the mapped area that are at equal elevations. The contour interval selected may be any value and unit. The shapes of the contour lines provide a representation of the shapes of hills and depressions. The lines themselves show the real elevations. Contour lines never touch one another. Contour lines never cross. Index contour lines mark the elevation for that line and are usually darker lines with numbers. Lines that are close together show a steep hill while wider spaced lines show a gentler slope. Rivers are shown by finding where the contour lines form Vs on the map. These Vs point upstream. Water always flows down hill. Looking at the shape of the contour lines and where they cross the stream will show you which way the water is flowing. Hachures are special marks that show hollows, or depressions.

Snowfall Amounts in New York State

While contour lines indicate relief along the horizontal, a profile indicates relief along the vertical. One can construct a profile by looking at a representative line on a relief map and plotting how the elevation changes along that line.

Objective:

The objective of this lab is to find points of equal value to illustrate a landform. The landform will then be arranged in profile to show an alternate view of the landform.

Essential Question:

____

____

Pre-Lab:

1. Define isoline: ____

   ____

2. Do isolines cross? ____

3. Do isolines touch? ____

Materials: • Isoline field • Ruler • Grid

Procedure:

Mark the isolines on the accompanying fields.
Remember, isolines never touch or cross.
Make sure the lines go to the edge of the map.

– Make the intervals in Field 1 a value of 2.

– Make the intervals in Field 2 a value of 20 feet.

Field 1

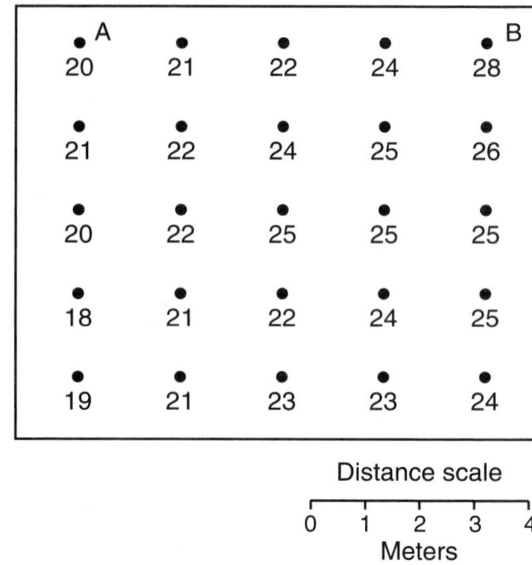

Field 2 **Elma Creek / Steeler Bay**

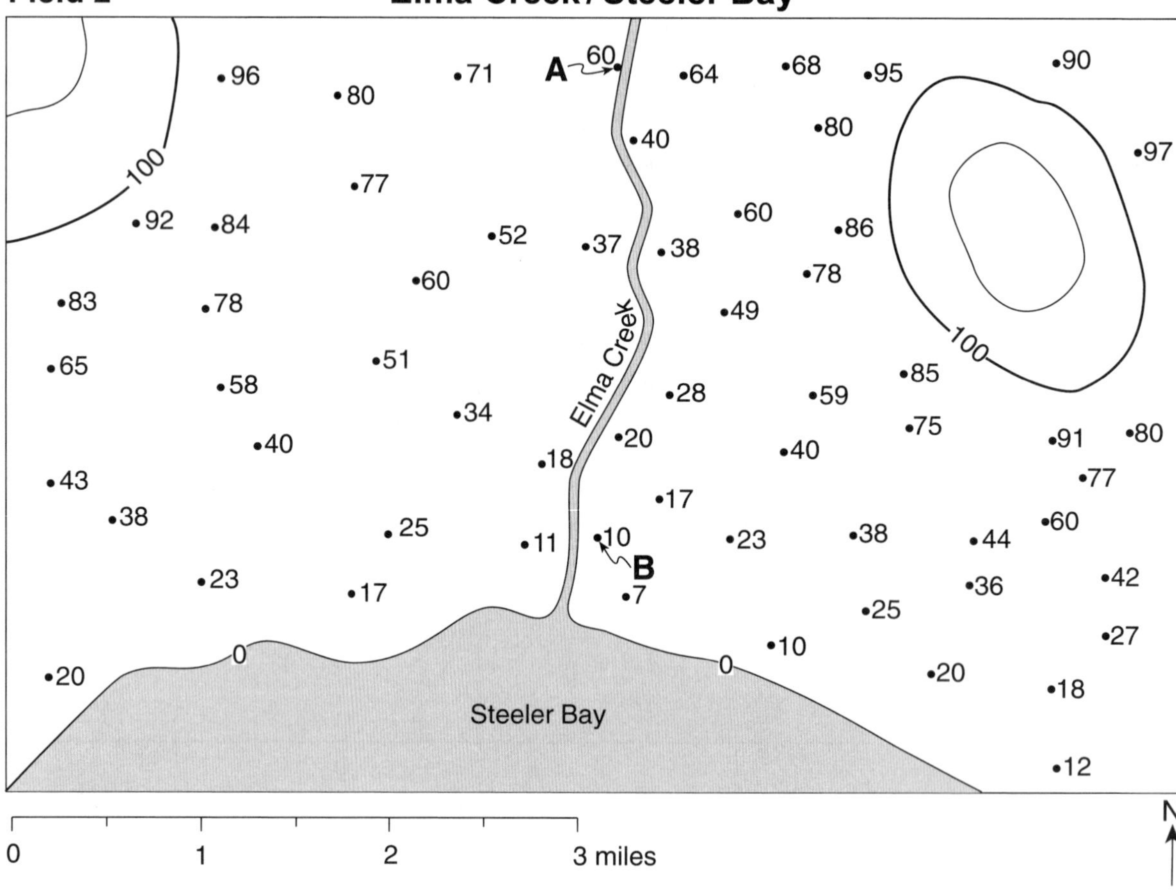

Now that you've created a field, use your skills to construct a profile of the contours from this map.

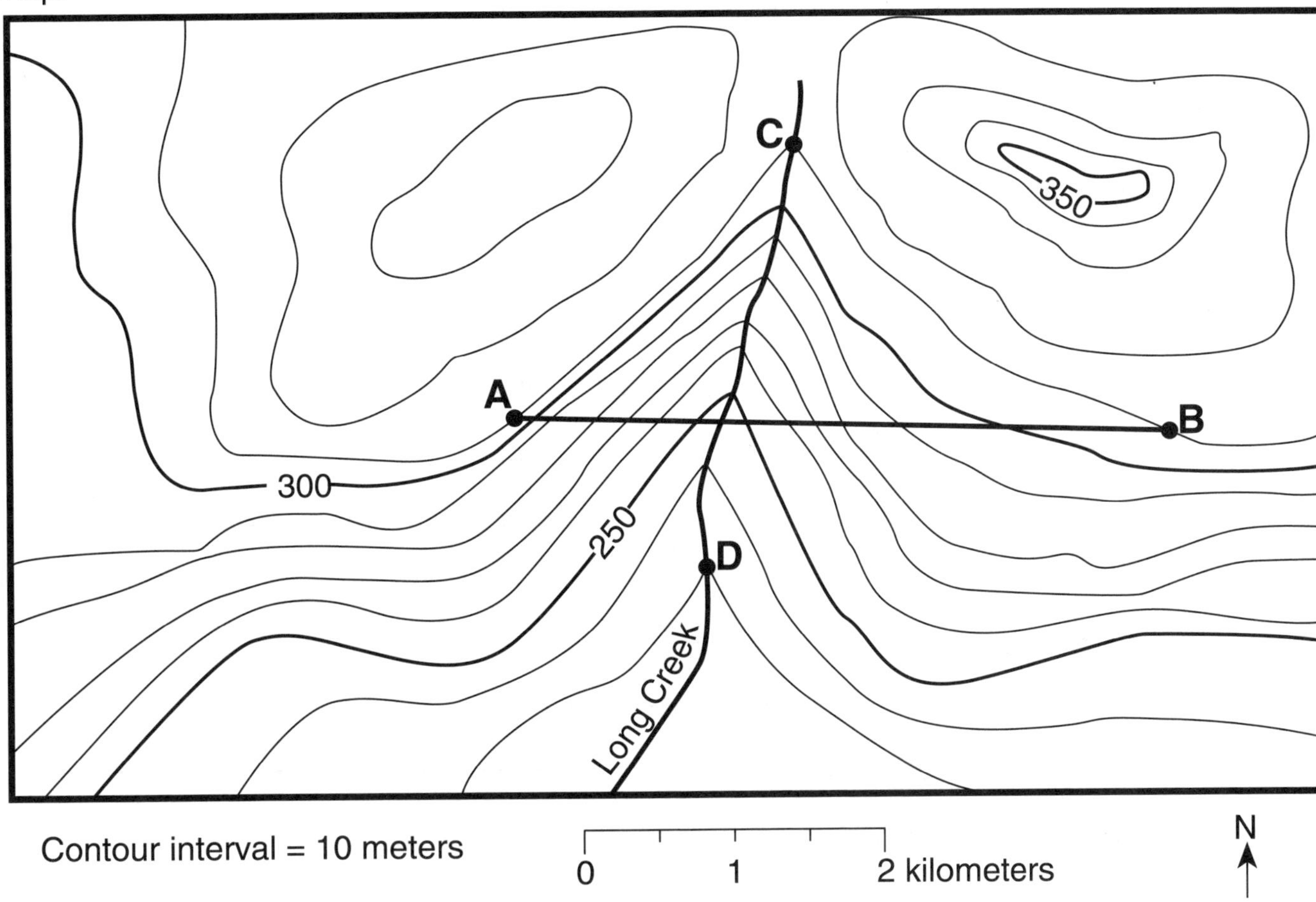

Profile Along Line AB

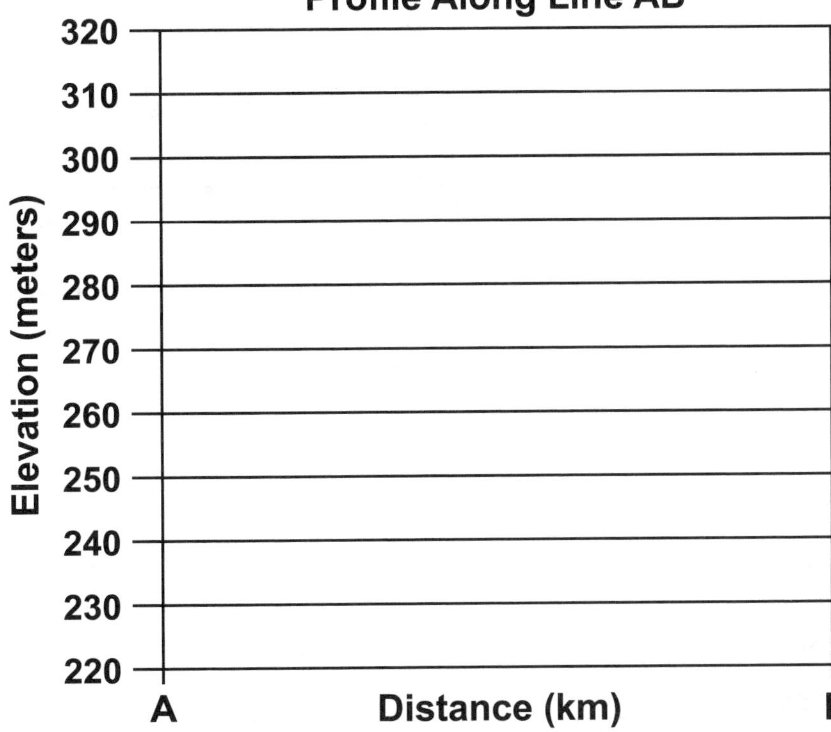

Observations:

1. What is being illustrated when isolines are close together? ______________________

____

2. Does every point from a field have to be included onto a contour line? Explain why or why not.

____

____

3. Referring back to the Field 2 map of Elma Creek / Steeler Bay, what direction does Elma Creek flow? ______________________

4. Explain how you know which direction Elma Creek flows. Make sure to use contour lines in your answer.

____

____

____

The following questions refer to the figures on the previous page.

5. What is the change in elevation from point C to point D? ______________________

6. What is the gradient from point C to point D? ______________________

7. What direction is Long Creek flowing? Support your answer. ______________________

____

8. Since you've completed your profile, which slope would be a better hike, east or west of Long Creek? ______________________

9. Referring to question number 8, explain your answer.

____

____

10. Look carefully at the landform, if you were building a shelter, where would you build? Explain why. ______________________

NAME ______________________________

4 3-D Topo Map

Cartography is the science of making maps. In the last project, contour maps were studied. In this investigation, contour planes will be made. Instead of simply drawing lines, entire field planes will be represented like a layer cake. In this investigation you will construct a three-dimensional model of an imaginary place.

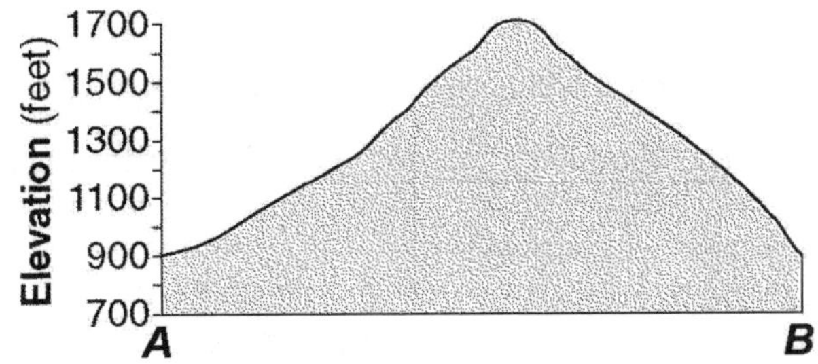

The maps on page 15 are different points of elevation on an oceanic island. The features are labeled and marked. Your ultimate goal is to create a three dimensional model of the landscape.

Objective: The objective of this lab is to physically illustrate topography of a known landform.

Essential Question:

____

____

Materials:
- Crayons or colored pencils
- 8 foam boards cut to 8" x 10"
- Scissors
- Glue

Rules of Contour Lines:

1. Every point on a contour line is a point of equal elevation.
2. Contour lines never touch.
3. Contour lines never cross.
4. Index contour lines are heavy lines that indicate elevation.
5. Closer lines show steeper gradient – farther lines show gentler gradient.
6. Vs point up-stream.
7. Hachures denote depression.

Procedure:

1. Study the rules of contour lines on page 13.
2. Cut out the map on page 15.
3. Coloring the Map:
   a) Color the area between the 0 index contour line and compass rose blue.
   b) Working toward Tall Mountain, color each successive contour interval in the order below: (don't forget about what hachures mean!)

(1) Dark Green	(4) Orange	(7) Brown
(2) Light Green	(5) Red	
(3) Yellow	(6) Purple	

4. Cut out each separate color field.
5. Glue the colored contour intervals onto a piece of foam board.
6. Cut out each foam board along the isoline, leaving the rest of the board intact so that the contours build up with each layer inside the dotted line.
7. Glue the layers together so that the contour lines align. Be sure to note hachures as well as index contour lines. A helpful hint would be to label the contour lines with a pencil. The contour intervals are measured in feet.

Observations:

1. What is the highest elevation of Tall Mountain? ____________________
2. What is the contour interval on the map? ____________________
3. Which slope has the steepest gradient? ____________________
4. What is the field value of each layer? (hint: it is a range) ____________________
5. What is the feature on the east side of the Tall Mountain?____________________
6. What is the highest possible elevation of the Small Hill? ____________________
7. What is the lowest possible elevation of the Lost Hollow? ____________________

____

9. What does the edge line of each layer have in common?

____

____

10. If you travel from the highest part of Tall Mountain to the lowest point of Lost Hollow, what is the change in field value?

____

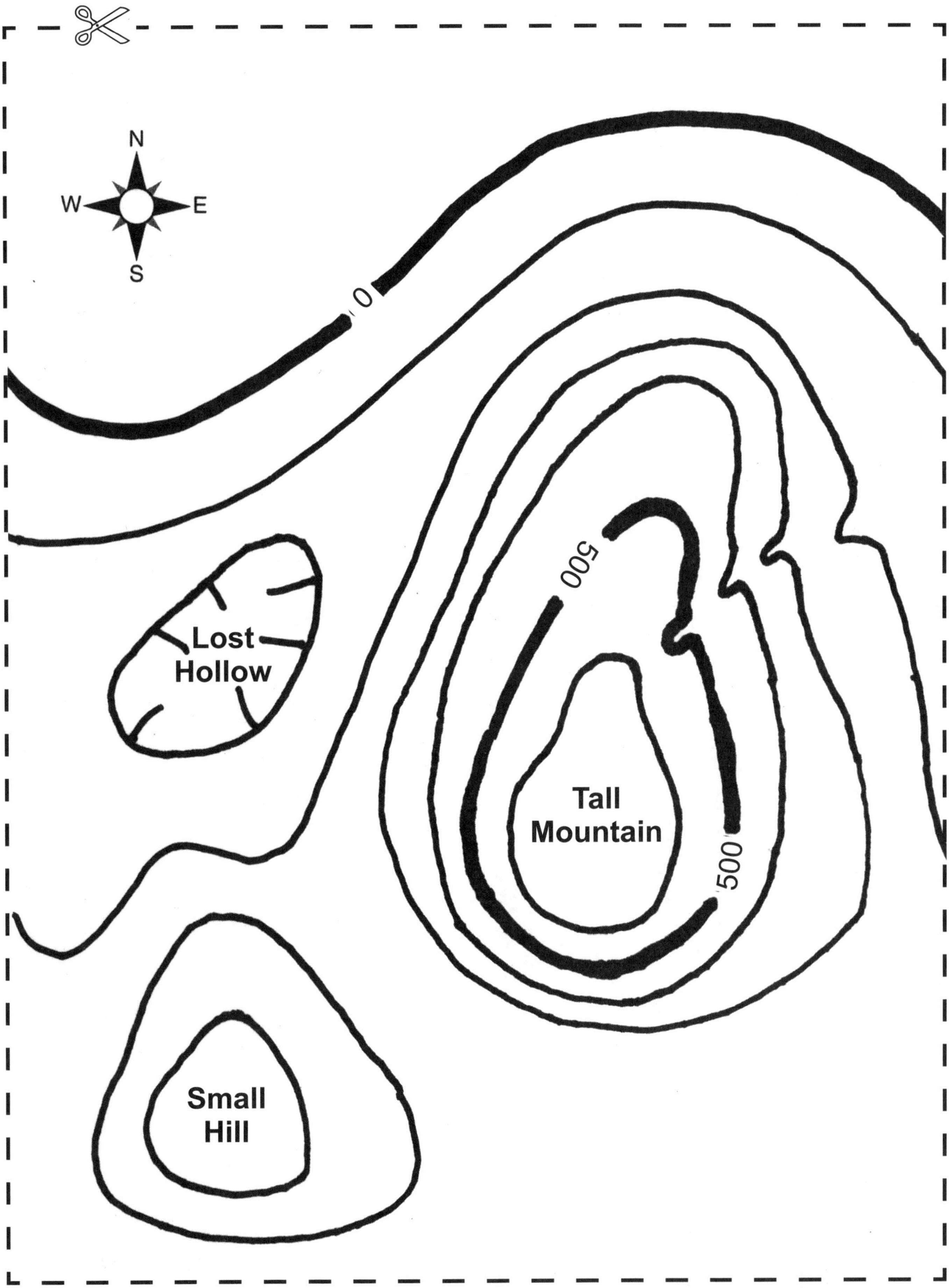
N
W
E
S
0
500
500
Lost Hollow
Tall Mountain
Small Hill

NAME ____________________

5 Latitude, Longitude and Time

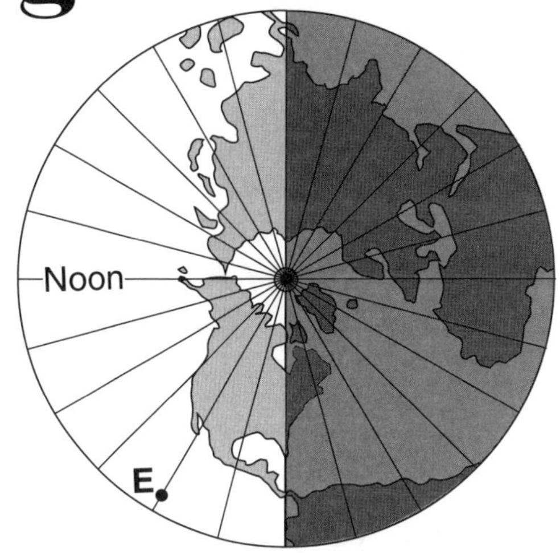

The Earth is divided into grids of latitude and longitude. Since the Earth is spherical, the grid cannot have equal units around the planet.

Lines of latitude indicate one's position north or south of the equator; they run east to west like rungs of a ladder. The lines of longitude meet at both poles and flare out around the equator. The astronomers at the Royal Observatory established Greenwich, England as the Prime Meridian, 0°. The lines of longitude start at the Prime Meridian and go 180° east and west around the globe were they meet at the International Dateline. The International Dateline is at 180°, the opposite side of the planet from the Prime Meridian. Times zones are areas around the planet where the time is standardized so that people living in that area would have a basic similarity in time. Time Zones do not always run along lines of longitude. Political boundaries exist to change the field.

Objective:

The objective of this investigation is to illustrate why time zones exist and variations in time zones.

Essential Question:

____

____

Hypothesis:

If ____________________

Then ____________________

Because ____________________

Variables:

Independent: ____________________

Dependent: ____________________

Material: • Ruler • Data chart • Colored pencils

Pre-Lab:

1. How many degrees are in a circle? ______________________________

2. How many hours are in a day? ______________________________

3. How many degrees does the earth rotate in one hour? (show your work here)

__________°

4. How many degrees does the sun appear to travel in an hour? __________°

Directions:

1. Look at the World Map on the next page. Starting from the prime meridian, carefully measure fifteen degrees. 15, 30, 45, etc... Using your colored pencils, shade in every other column with a color of your choice.
2. Assuming it is noon at the Prime Meridian, write the times along the longitudinal lines in the World Map. 11:00 am and noon have been completed for you.
3. Plot each city, from the chart below, onto the World Map (page 19). Use the letter in front of each city to mark its location on the World Map.
4. Complete the Time column of the Cities of The World chart below. London, England has been done for you.

CITIES OF THE WORLD

City	Latitude	Longitude	Time
A. London, England	51° N	0°	Noon
B. Reykjavík, Iceland	64° N	21° W	
C. Rio de Janeiro, Brazil	22° S	43° W	
D. Cape Town, South Africa	33° S	18° E	
E. New York City, United States	40° N	74° W	
F. Sydney, Australia	33° S	151° E	
G. Yucatán, Mexico	20° N	89° W	
H. Christchurch, New Zealand	43° S	172° E	
I. Vancouver, Canada	49° N	123° W	
J. Puerto Williams, Chile	54° S	67° W	

WORLD MAP

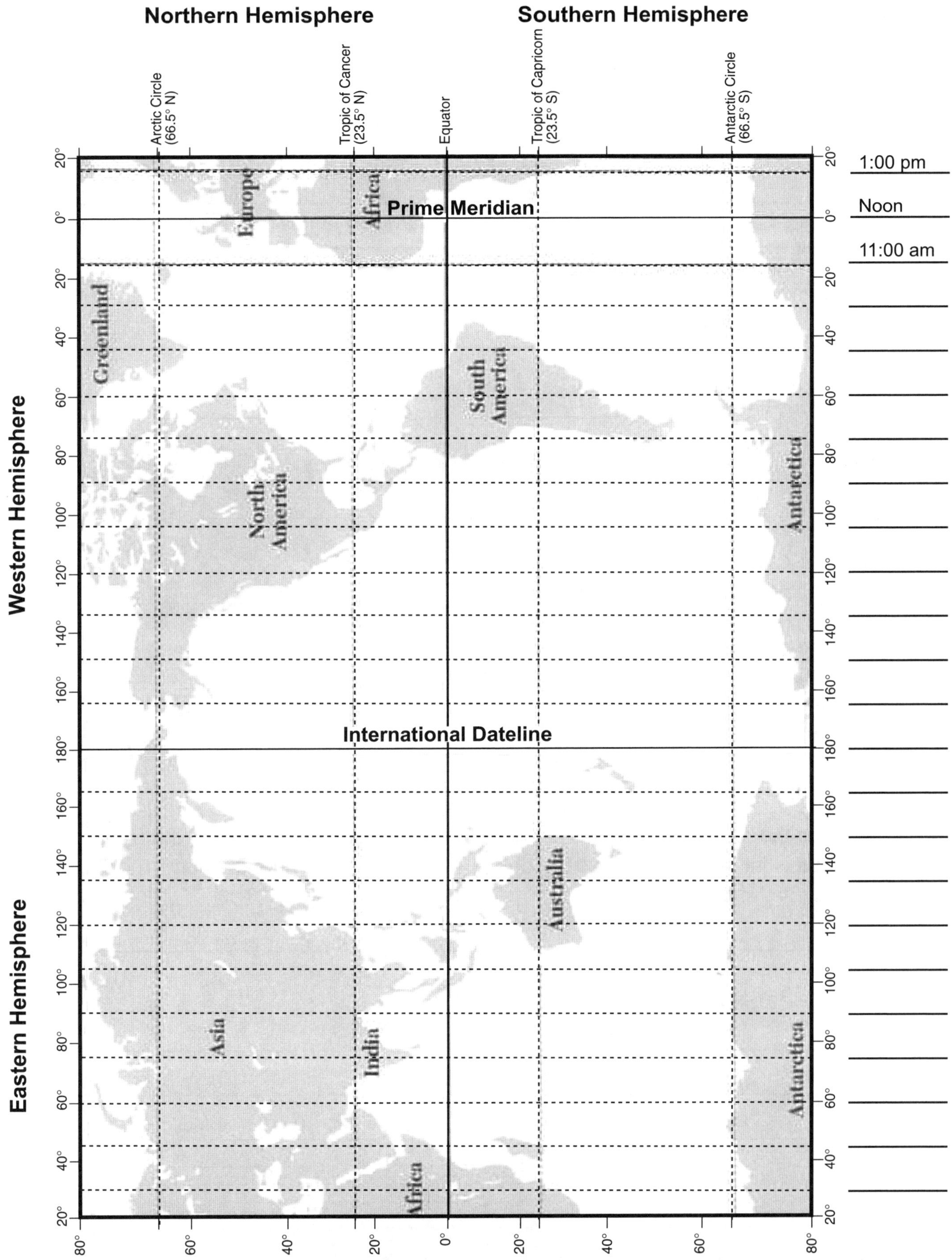

Northern Hemisphere
Southern Hemisphere
Western Hemisphere
Eastern Hemisphere
Arctic Circle (66.5° N)
Tropic of Cancer (23.5° N)
Equator
Tropic of Capricorn (23.5° S)
Antarctic Circle (66.5° S)
Prime Meridian
International Dateline
Europe
Africa
Greenland
North America
South America
Antarctica
Australia
Asia
India
1:00 pm
Noon
11:00 am

Observations:

1. What direction does the Sun rise? ______________________

2. Thinking about question 1, which location would reach noon first? The eastern city or the western city?

____

____

3. About how wide do you think a time zone should be?

____

4. What is your time zone? ______________________

5. Which city will see noon first, Rio de Janeiro or New York City?

____

6. What is the calculated difference in time between Rio de Janeiro and New York City? ______________________

7. What is the difference in degrees (in longitude) between Yucatán, Mexico and London, England? ______________________

8. Which city would see noon first, Cape Town or London?

____

9. At midnight the residents of Tokyo, Japan (35° N 139° E) have just rung in the New Year. Based on this information, what is the date and time for the residents of Los Angeles, California (34° N 118° W)?

____

10. Why do we have time zones?

____

____

____

____

____

6 United States Time Zones

NAME ________________________

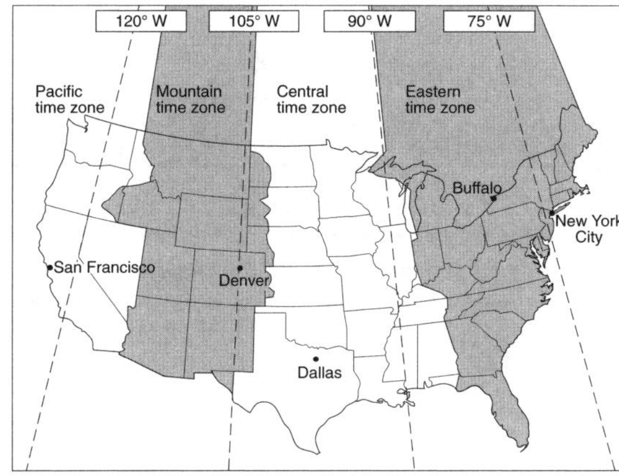

Often when watching television we see a show starting at 9 o'clock eastern, 8 central. We've heard it so much, we don't even pay attention to it. When our eastern sports teams play on the west coast, we see it is daylight on the televisions while it's nighttime here.

The United States is an expansive country with states and territories spanning the planet. The eastern most town in the contiguous United States is Lebuc, Maine with the terrestrial coordinates of 45° N and 67° W. The western most town is Ozette, Washington with terrestrial coordinates of 48° N 124° W. Key West, Florida claims many landmarks to be the southern-most point in the United States. When Hawaii, Alaska, and the territories are included, we can cross the International Date Line. Amundsen-Scott South Pole Station is a research station with United States jurisdiction in Antarctica. It's terrestrial coordinates are 89° 99′ S and 0° 0′ E.

Objective:

The objective of this investigation is to illustrate why we have time zones in the United States and to find the extremes of the United States as well as within New York State.

Essential Question:

____

____

Hypothesis:

If ____

Then ____

Because ____

Variables:

Independent: ____

Dependent: ____

Materials: • Calculator • Maps • Globe • External protractor

Pre-lab: Locate the cities on the map below. Record the cities time zones in the spaces provided.

City	Latitude & Longitude	Time Zone
New York, New York	40° 43′ N 74° 00′ W	
Dallas, Texas	32° 78′ N 98° 80′ W	
Denver, Colorado	39° 44′ N 104° 59′ W	
San Francisco, California	37° 46′ N 122° 26′ W	

Assuming it is 9:00 a.m., in the central time zone, fill in the correct time for the remaining time zones.

Time Zones

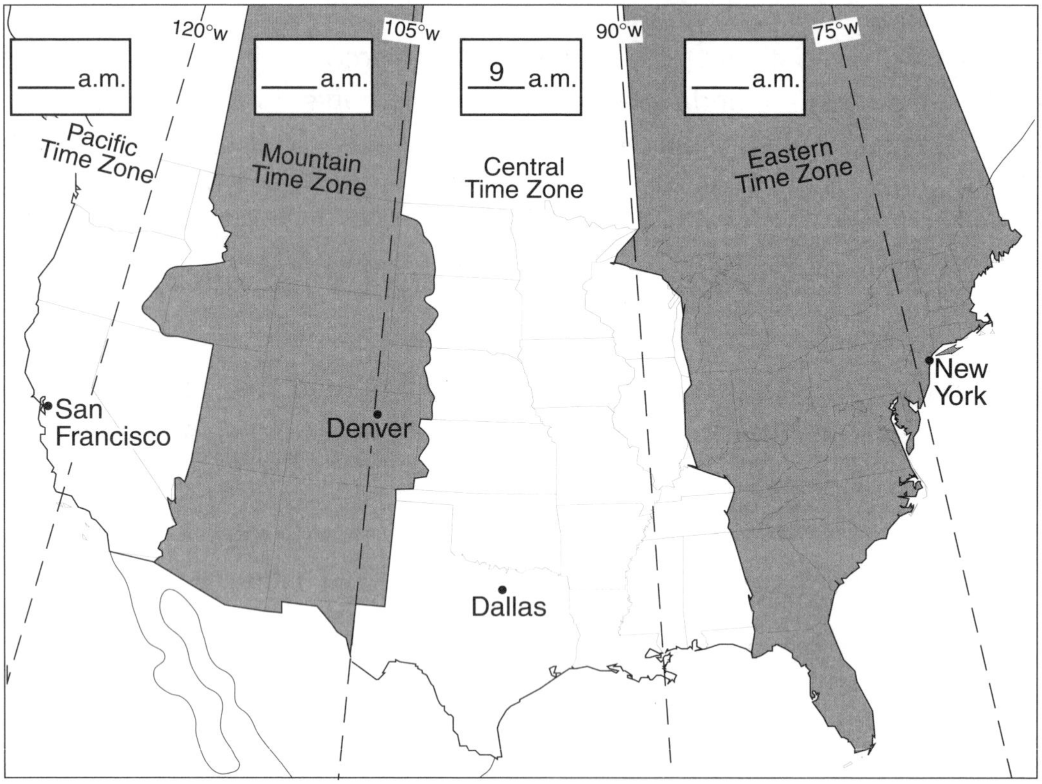

Observations:

1. Which is the northern most city of the four listed above? ______________________________

2. Approximately, about how far apart are the four cities by longitude? ____________________

3. The Texas Rangers are playing the San Francisco Giants in California with a starting time of 8:00 p.m.. What time would residents in Dallas expect to see the game begin on their televisions?

____

Directions:

1. Using your globe, take a piece of masking tape and place it on the Equator starting at 150°W and, going from left to right, continue to 50°W.

2. Please find the following locations on the globe:

   A. Barrow, AK 71°N 156°W
   B. NW Angle, MN 49°N 95°W
   C. Ka Lae, HI 18°N 155°W
   D. Key West, FL 24°N 81°W
   D. Wake Island 19°N 166°E
   E. Lubec, ME 44°N 66°W
   F. Amatignak Island, AK 51°N 179°W
   G. Ozette, WA 48°N 124°W

3. When these points are located, trace with your finger the longitude line southward to the Equator.

4. Write the letters representing each city on the tape on the appropriate longitude. Example: A. represents Barrow AK. It would be written on the tape closest to 156°W.

5. Use an external protractor to complete the following

Observations:

1. Using your external protractor, what is the difference in longitude between Wake Island and Ke Lae, HI? What is the time difference?

   ____

2. Look at Barrow, AK and Ka Lae, HI. What is the difference in time zones?

   ____

3. What is the time difference between Ozette, WA and Amitignak Island, AK?

   ____

4. What is the difference between the longitude from Lubec, ME and Key West, FL?

   ____

5. It is Thanksgiving Day in Lubec, ME and your clock strikes noon. What is the time difference when residents of Key West, FL hear their own noon chimes?

   ____

Directions: Refer to Generalized Bedrock Geology of New York State map below to answer questions 6 through 10.

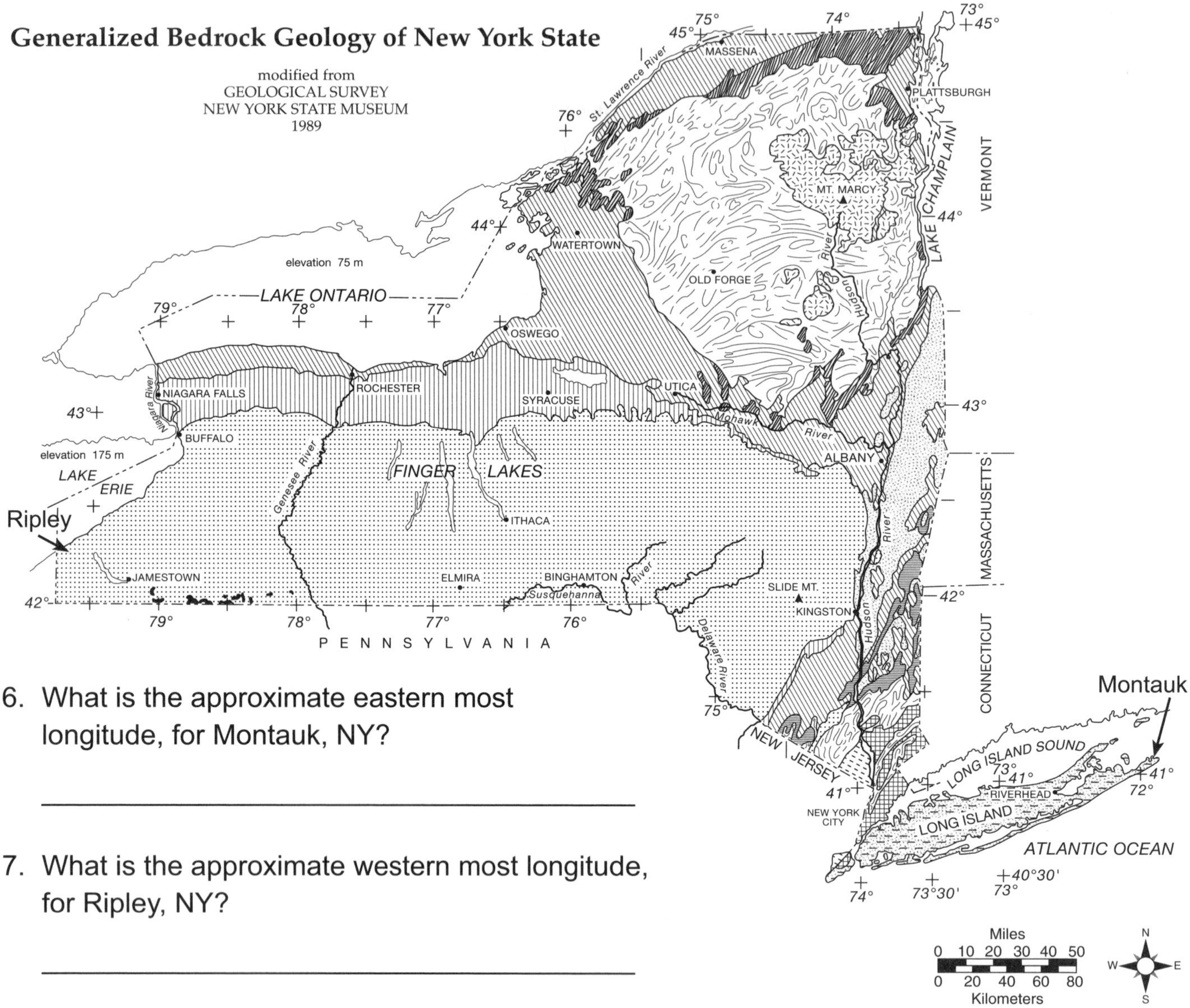

6. What is the approximate eastern most longitude, for Montauk, NY?

____

7. What is the approximate western most longitude, for Ripley, NY?

____

8. What is the approximate difference in degrees (in longitude) between these two points?

____

9. Which city will see the sunrise first and how much time will pass before the second city sees the sunrise?

____

10. Students in both Ripley and Montauk will be taking the Earth Science Regents exam in June at 9:15 am. Explain why they both should be present and ready to take the exam at 9:15 am.

____

7 Crystal Growth

NAME ______________________

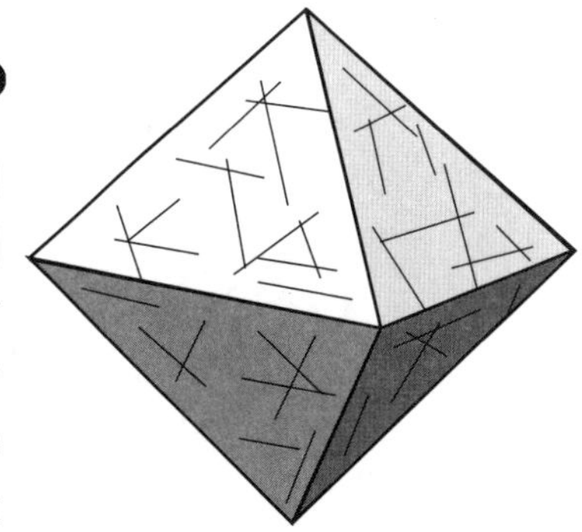

Crystals are the basic building blocks of minerals. To understand the way in which minerals are formed, the structure of crystals and crystal growth must be understood. When crystal formation, or growth, is understood the properties of minerals are more easily understood. Minerals are naturally occurring substances that are not organic in origin. Non-organic means that the materials that make up minerals were never alive. Minerals vary in their properties because of the internal arrangement of atoms. The internal arrangement of atoms is how the atoms are attached to each other in the mineral lattice.

Examples of common crystal formations

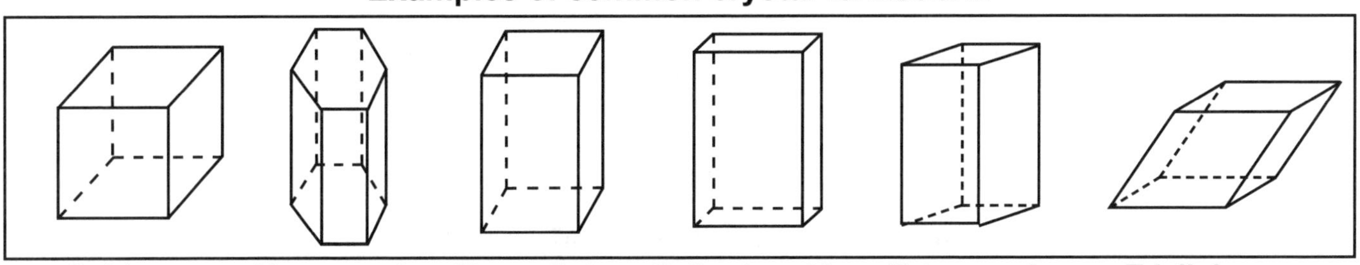

Cubic **Hexagonal** **Monoclinic** **Orthorhombic** **Tetragonal** **Triclinic**

Objective:

The objective of this lab is to grow crystals in order to determine the characteristics of those crystals.

Essential Question:

____

____

Hypothesis:

If____

Then____

Because____

Variables:

Independent:____

Dependent:____

Materials:

- Heat resistant glass jar
- Crystal shape chart
- 2 popsicle sticks
- Salts*
- Heating plate
- Magnifying glass
- String
- Paper clip
- Beaker
- Water

* Borax, Baking Soda, Table Salt, Sea Salt, Halite or powdered soap

Procedure:

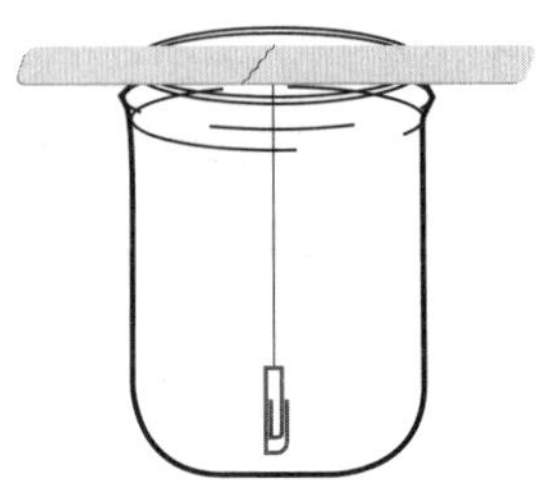

1. Tie one end of the string to the paper clip.
2. Place one popsicle stick across the top of the jar.
3. Carefully lower the paper clip into the jar so that the paper clip is just off the bottom.
4. Mark on the string where it meets the popsicle stick.
5. Remove the string and paper clip from the jar.
6. Tie the string to the popsicle stick using your mark as a guide.
7. Replace the paperclip and string back into the jar to make sure the paper clip doesn't touch the bottom of the jar. Adjust the length of the string if necessary.
8. Bring the water to a boil and then pour it into the jar approximately ¾ full.
9. Carefully add salt* while stirring with the other popsicle stick.
10. Stir until water becomes cloudy and some salts settle to the bottom.
11. Put the string / paperclip assembly back into jar.
12. The jar should be put in a place where it won't be disturbed for several days.

* Borax, Baking Soda, Table Salt, Sea Salt, Halite or powdered soap

Observations: **- After 3 days-** **- Draw Here -**

1. What do you observe on the string?

____

____

2. Describe the shapes of the objects.

____

____

3. Classify what you observe based on the crystal shape chart.

____

____

4. How are the objects grouped; chains, clusters?

____

____

5. What do you think will happen after 4 more days?

____

____

- After 7 days- **- Draw Here -**

6. What difference do you observe on the string?

____

____

7. What is the crystal that solidified out of the solution called?

____

____

8. Classify what you observe based on the crystal shape chart.

____

____

9. What do you think will happen if left in longer?

____

____

10. Infer: What does this experiment tell you about crystal growth and time?

____

____

____

____

8 Minerals

NAME ______________________

Minerals are any naturally occurring chemical elements or compounds. In geology, chemical elements and compounds have been formed through inorganic processes. There are over 3000 minerals known on planet Earth; they differ in their chemical and physical attributes. They are classified primarily by chemical composition, the type of crystal they are and their hardness. Appearance is also very important in the determination of minerals. Minerals are solid substances that are not man-made. Most of the rocks forming the Earth's crust consist of minerals.

Moh's Hardness Scale

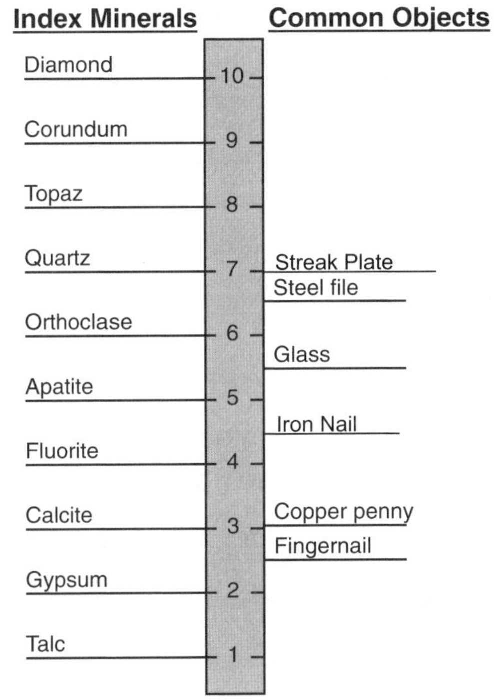

Objective:

The objective of this lab is to identify minerals using the Properties of Common Minerals chart of the Earth Science Reference Tables.

Essential Question:

____

____

****CAUTION! THIS LAB EMPLOYS THE USE OF AN ACID, USE EXTREME CAUTION.****

Material:

- Minerals
- Moh's Scale
- Iron nail
- Steel file
- Copper
- Glass plate
- Goggles
- Streak plate
- HCl
- Properties of Common Minerals chart (page 218)

Procedure:

Use the Properties of Common Minerals chart (page 218), and the Moh's Hardness Scale to identify the minerals in your kit. Answer all questions for each mineral.

Mineral 1

1. Luster: Metallic – Non metallic (circle one)
2. Cleavage or fracture?__________________________
3. Color: ____________________________________
4. Hardness: ________________________________

   The mineral can scratch: (Starting with the copper penny, if it is scratched, move onto the next item. Continue until you discover the item the mineral can not scratch.)

   copper penny - iron nail - glass plate - steel file - streak plate

   The hardness must be more than: _______________

   The mineral is scratched by: (Starting with the fingernail, if it does not scratch the mineral, move onto the next item. Continue until you discover the item that can scratch the mineral.)

   fingernail - copper penny - iron nail - glass plate - steel file - streak plate

   The hardness must be less than: ________________

5. Streak color:__________________________________
6. Does it bubble with acid?___________ This mineral is:______________________________

Mineral 2

1. Luster: Metallic – Non metallic (circle one)
2. Cleavage or fracture?__________________________
3. Color: ____________________________________
4. Hardness: ________________________________

   The mineral can scratch: (Starting with the copper penny, if it is scratched, move onto the next item. Continue until you discover the item the mineral can not scratch.)

   copper penny - iron nail - glass plate - steel file - streak plate

   The hardness must be more than: _______________

   The mineral is scratched by: (Starting with the fingernail, if it does not scratch the mineral, move onto the next item. Continue until you discover the item that can scratch the mineral.)

   fingernail - copper penny - iron nail - glass plate - steel file - streak plate

   The hardness must be less than: ________________

5. Streak color:__________________________________
6. Does it bubble with acid?___________ This mineral is:______________________________

Mineral 3

1. Luster: Metallic – Non metallic (circle one)
2. Cleavage or fracture?______________________________
3. Color: ______________________________
4. Hardness: ______________________________

   The mineral can scratch: (Starting with the copper penny, if it is scratched, move onto the next item. Continue until you discover the item the mineral can not scratch.)

   copper penny - iron nail - glass plate - steel file - streak plate

   The hardness must be more than: ______________

   The mineral is scratched by: (Starting with the fingernail, if it does not scratch the mineral, move onto the next item. Continue until you discover the item that can scratch the mineral.)

   fingernail - copper penny - iron nail - glass plate - steel file - streak plate

   The hardness must be less than: ______________
5. Streak color:______________________________
6. Does it bubble with acid?__________ This mineral is:______________________________

Mineral 4

1. Luster: Metallic – Non metallic (circle one)
2. Cleavage or fracture?______________________________
3. Color: ______________________________
4. Hardness: ______________________________

   The mineral can scratch: (Starting with the copper penny, if it is scratched, move onto the next item. Continue until you discover the item the mineral can not scratch.)

   copper penny - iron nail - glass plate - steel file - streak plate

   The hardness must be more than: ______________

   The mineral is scratched by: (Starting with the fingernail, if it does not scratch the mineral, move onto the next item. Continue until you discover the item that can scratch the mineral.)

   fingernail - copper penny - iron nail - glass plate - steel file - streak plate

   The hardness must be less than: ______________
5. Streak color:______________________________
6. Does it bubble with acid?__________ This mineral is:______________________________

Mineral 5

1. Luster: Metallic – Non metallic (circle one)
2. Cleavage or fracture?____________________________
3. Color: ______________________________________
4. Hardness: ____________________________________

   The mineral can scratch: (Starting with the copper penny, if it is scratched, move onto the next item. Continue until you discover the item the mineral can not scratch.)

   copper penny - iron nail - glass plate - steel file - streak plate

   The hardness must be more than: ________________

   The mineral is scratched by: (Starting with the fingernail, if it does not scratch the mineral, move onto the next item. Continue until you discover the item that can scratch the mineral.)

   fingernail - copper penny - iron nail - glass plate - steel file - streak plate

   The hardness must be less than: __________________
5. Streak color:____________________________________
6. Does it bubble with acid?____________ This mineral is:______________________________

Observations:

1. What was the most reliable characteristic of the mineral to identify? _____________________

   ____
2. Can color be a definitive factor in determining a mineral? ____________________________
3. In referring to question 2, why? ____

   ____

   ____

   ____
4. Why did you scratch the mineral with different items?

   ____

   ____

   ____

   ____

5. What differences did you see in the different minerals? ____________________

____

____

____

6. Does the color of the streak match the color of the mineral? ____________________

____

____

____

Conclusions:

7. Why does a mineral get scratched for identification? ____________________

____

____

____

____

8. It is said that minerals cannot be identified only by color. Explain why and use a mineral as an example.

____

____

____

____

9. Both graphite and diamond are elemental minerals, meaning that they each consist only of one element; carbon. Graphite has a Moh's Hardness of 1 – 2 while a diamond has a Moh's Hardness of 10. What can be inferred about the differences of graphite and diamond?

____

____

____

____

____

10. Pretending your lab partner was absent, write a letter to your lab partner explaining how to identify minerals. Use your own words.

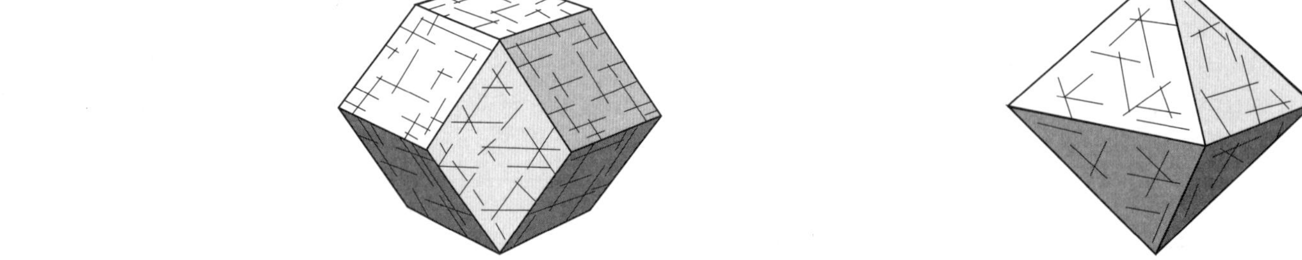

NAME ____________________

9 Igneous Rock Identification

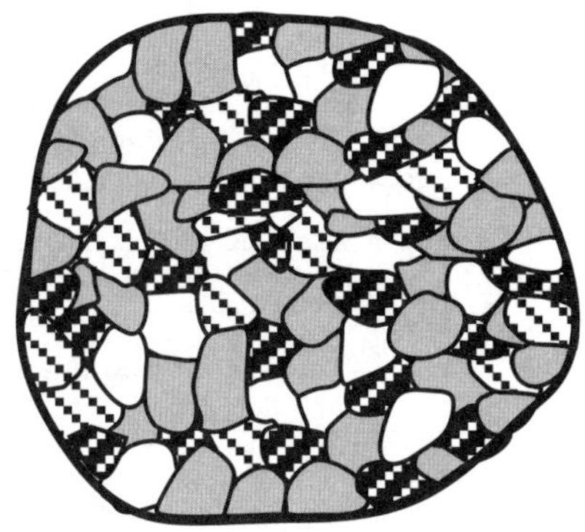

Igneous rocks are rocks that form when lava or magma cools. Lava and magma are both molten minerals. Lava is magma that has come out of the volcano.

Rock that is formed when lava cools is called extrusive or volcanic. Lava on the exterior of the volcano is exposed to the air or water (underwater volcanoes) and cools quickly. These rocks sometimes don't even have time to form crystals; this produces a glassy texture. Sometimes gas gets trapped in the rock, giving the rock an appearance of being spongy or vesicular texture. Rock that is formed when magma cools is called intrusive or plutonic.

Magma is in the interior of the volcano. The rock surrounding the volcano is an excellent insulator, making the magma cool very slowly. The mineral crystals can grow very large.

Geologists also classify igneous rocks based on the minerals in the rock. Light colored and less dense rocks are called felsic. Denser darker colored rocks are called mafic.

Since the internal arrangement of atoms is the critical factor in determining the properties of minerals, how the magma cools determines what type of mineral forms.

Objective:

The objective of this lab is to identify the igneous rocks using the Earth Science Reference Tables and by observing the characteristics of the rock.

Essential Question:

____

____

Material:

- Igneous rocks kit
- Metric ruler
- Magnifying lens
- Scheme for Igneous Rock Identification chart found on page 36

Procedure:

Use the Scheme for Igneous Rock Identification chart below, magnifying lens, and metric ruler to identify the rocks from your kit. Map out your findings using the dichotomous keys.

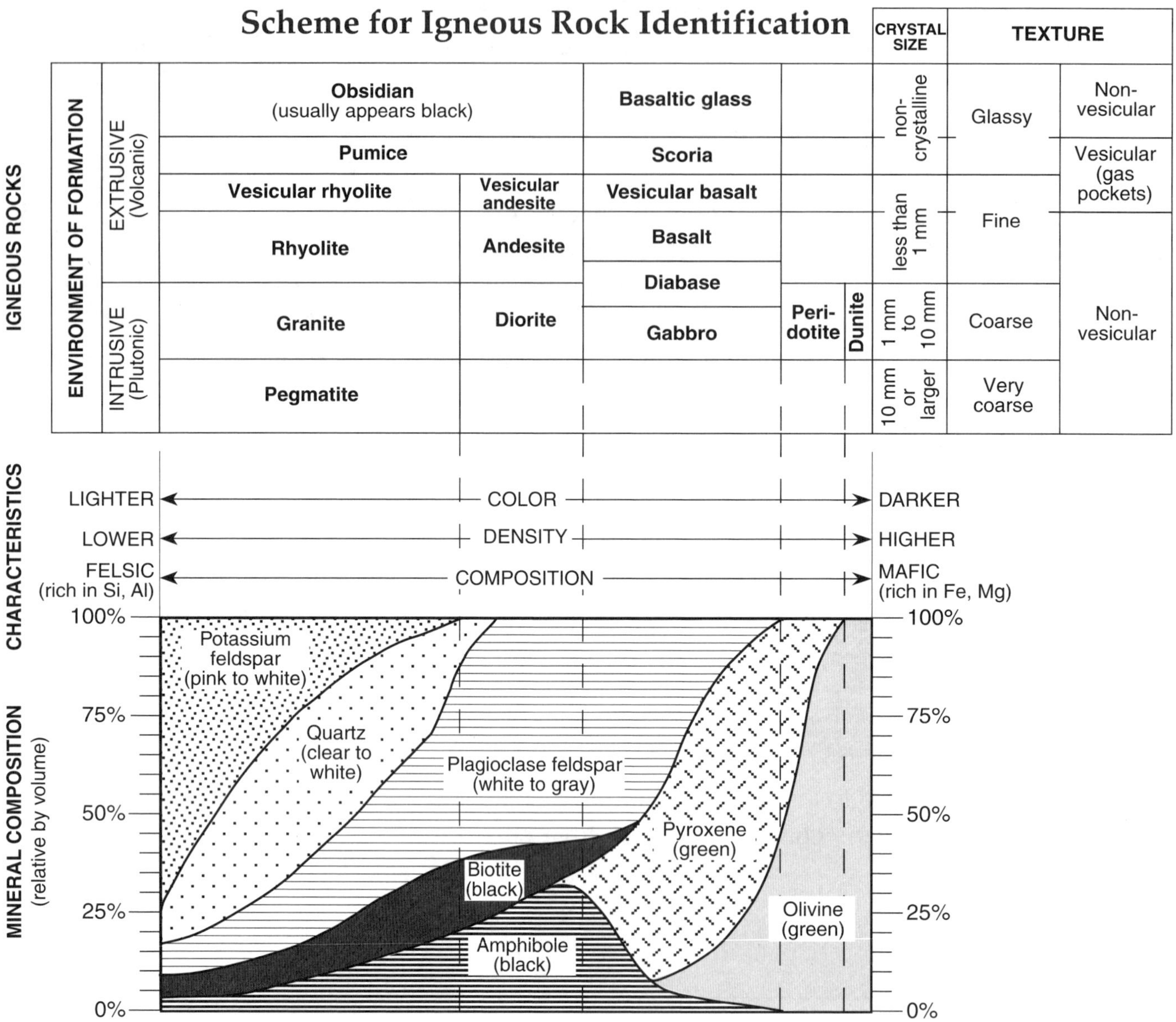

Rock #1

1. What is the crystal texture? (circle one) Glassy – Vesicular – Non-Vesicular
2. The grain size is: non-crystalline / < 1mm / 1mm – 10mm / 10mm or larger
3. Looking at questions 1 and 2 the rock is (extrusive / intrusive).
4. The rock is (light colored pink to light grey / dark colored dark grey to green).
5. Looking at question number 4, the rock is (felsic / mafic).
6. Based on the answers above, the rock is ______________________.

Rock #2

1. What is the crystal texture? (circle one) Glassy – Vesicular – Non-Vesicular
2. The grain size is: non-crystalline / < 1mm / 1mm – 10mm / 10mm or larger
3. Looking at questions 1 and 2 the rock is (extrusive / intrusive).
4. The rock is (light colored pink to light grey / dark colored dark grey to green).
5. Looking at question number 4, the rock is (felsic / mafic).
6. Based on the answers above, the rock is ________________________.

Rock #3

1. What is the crystal texture? (circle one) Glassy – Vesicular – Non-Vesicular
2. The grain size is: non-crystalline / < 1mm / 1mm – 10mm / 10mm or larger
3. Looking at questions 1 and 2 the rock is (extrusive / intrusive).
4. The rock is (light colored pink to light grey / dark colored dark grey to green).
5. Looking at question number 4, the rock is (felsic / mafic).
6. Based on the answers above, the rock is ________________________.

Rock #4

1. What is the crystal texture? (circle one) Glassy – Vesicular – Non-Vesicular
2. The grain size is: non-crystalline / < 1mm / 1mm – 10mm / 10mm or larger
3. Looking at questions 1 and 2 the rock is (extrusive / intrusive).
4. The rock is (light colored pink to light grey / dark colored dark grey to green).
5. Looking at question number 4, the rock is (felsic / mafic).
6. Based on the answers above, the rock is ________________________.

Rock #5

1. What is the crystal texture? (circle one) Glassy – Vesicular – Non-Vesicular
2. The grain size is: non-crystalline / < 1mm / 1mm – 10mm / 10mm or larger
3. Looking at questions 1 and 2 the rock is (extrusive / intrusive).
4. The rock is (light colored pink to light grey / dark colored dark grey to green).
5. Looking at question number 4, the rock is (felsic / mafic).
6. Based on the answers above, the rock is ________________________.

Observations:

1. What are the differences between granite and rhyolite?

2. What are the differences between granite and gabbro?

3. What does vesicular mean?

4. Are mafic rocks more or less dense than felsic rocks?

5. What are the three main differences between felsic and mafic rocks?

6. What is the mineral content of diorite?

____

____

7. What is the mineral content of vesicular rhyolite?

____

____

8. How is dunite unique on the Scheme for Igneous Rock Identification chart (page 36)?

____

____

____

9. What are the elements that make mafic rocks dense?

____

____

____

10. How does knowing the rocks' properties help you to identify rocks?

____

____

____

____

____

____

NAME ______________________

10 Sedimentary Rock Identification

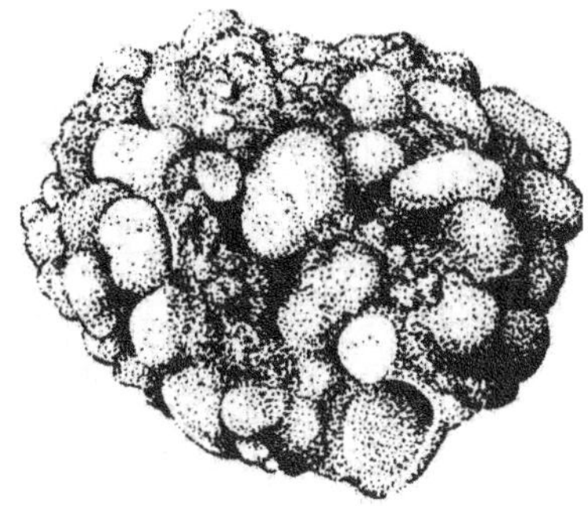

Sedimentary rock is formed from the cementation and compaction of loose material called sediment. Sedimentary rocks are classified as inorganic (clastic) or organic (crystalline and bioclastic). Clastic rocks are rocks formed by debris; broken rock fragments that vary in size. The size can be very large particles, 100.0 cm to very small, 0.00001 cm in size.

Most fossils are found in sedimentary rocks because the processes that form igneous and metamorphic rocks prevent fossilization or would likely destroy fossils.

Sedimentary rock forms when layers of sediment are compacted and cemented together. Sedimentary rock constitutes nearly 75% of the exposed rock on the earth's surface. Clastic rocks can be determined by appearance and grain size. Organic rocks are rocks formed from once living things.

Chemical sedimentary rocks tend to be monomineralic, having only one mineral, because of the processes that formed it.

Objective:

The objective of this lab is to identify sedimentary rocks using the Earth Science Reference Table.

Essential Question:

____

____

Materials:

- 5 Samples of sedimentary rock
- Scheme for Sedimentary Rock Identification chart on page 42
- Magnifying lens

Scheme for Sedimentary Rock Identification

INORGANIC LAND-DERIVED SEDIMENTARY ROCKS					
TEXTURE	**GRAIN SIZE**	**COMPOSITION**	**COMMENTS**	**ROCK NAME**	**MAP SYMBOL**
Clastic (fragmental)	Pebbles, cobbles, and/or boulders embedded in sand, silt, and/or clay	Mostly quartz, feldspar, and clay minerals; may contain fragments of other rocks and minerals	Rounded fragments	**Conglomerate**	
			Angular fragments	**Breccia**	
	Sand (0.006 to 0.2 cm)		Fine to coarse	**Sandstone**	
	Silt (0.0004 to 0.006 cm)		Very fine grain	**Siltstone**	
	Clay (less than 0.0004 cm)		Compact; may split easily	**Shale**	
CHEMICALLY AND/OR ORGANICALLY FORMED SEDIMENTARY ROCKS					
TEXTURE	**GRAIN SIZE**	**COMPOSITION**	**COMMENTS**	**ROCK NAME**	**MAP SYMBOL**
Crystalline	Fine to coarse crystals	Halite	Crystals from chemical precipitates and evaporites	**Rock salt**	
		Gypsum		**Rock gypsum**	
		Dolomite		**Dolostone**	
Crystalline or bioclastic	Microscopic to very coarse	Calcite	Precipitates of biologic origin or cemented shell fragments	**Limestone**	
Bioclastic		Carbon	Compacted plant remains	**Bituminous coal**	

Directions:

Use your magnifying lens and the chart above to determine the following rocks. Be sure not to mix up the rock samples.

Rock #1

1. Is the rock dark in color, and feels like wood? ________ If yes, you have coal, go to 9. If no, go to 2.
2. Does the rock bubble with acid? _______ If yes, you have limestone, go to 9. If no, go to 3.
3. Is the rock clear with a definite cube like shape? ________ If yes, you have rock salt, go to 9. If no, go to 4.
4. Can you scratch the rock with your fingernail? ________ If yes, you have rock gypsum, go to 9. If no, go to 5.
5. Is the rock very smooth with no visible grains? ________ If yes, you have shale, go to 9. If no, go to 6.
6. Does the rock have a gritty feel but the grains are difficult to see? ________ If yes, you have siltstone, go to 9. If no, go to 7.
7. Does the rock have a texture of sandpaper with visible grains ________ if yes, you have sandstone, go to 9. If no, go to 8.
8. Does rock have clearly visible particles, either rounded or angular? _______ You have either conglomerate or breccia, check with reference table, then go to 9.
9. The rock is ______________________________

Rock # 2

1. Is the rock dark in color, and feels like wood? ________ If yes, you have coal, go to 9. If no, go to 2.
2. Does the rock bubble with acid? ________ If yes, you have limestone, go to 9. If no, go to 3.
3. Is the rock clear with a definite cube like shape? ________ If yes, you have rock salt, go to 9. If no, go to 4.
4. Can you scratch the rock with your fingernail? ________ If yes, you have rock gypsum, go to 9. If no, go to 5.
5. Is the rock very smooth with no visible grains? ________ If yes, you have shale, go to 9. If no, go to 6.
6. Does the rock have a gritty feel but the grains are difficult to see? ________ If yes, you have siltstone, go to 9. If no, go to 7.
7. Does the rock have a texture of sandpaper with visible grains ________ if yes, you have sandstone, go to 9. If no, go to 8.
8. Does rock have clearly visible particles, either rounded or angular? ________ You have either conglomerate or breccia, check with reference table, then go to 9.
9. The rock is ____________________________

Rock # 3

1. Is the rock dark in color, and feels like wood? ________ If yes, you have coal, go to 9. If no, go to 2.
2. Does the rock bubble with acid? ________ If yes, you have limestone, go to 9. If no, go to 3.
3. Is the rock clear with a definite cube like shape? ________ If yes, you have rock salt, go to 9. If no, go to 4.
4. Can you scratch the rock with your fingernail? ________ If yes, you have rock gypsum, go to 9. If no, go to 5.
5. Is the rock very smooth with no visible grains? ________ If yes, you have shale, go to 9. If no, go to 6.
6. Does the rock have a gritty feel but the grains are difficult to see? ________ If yes, you have siltstone, go to 9. If no, go to 7.
7. Does the rock have a texture of sandpaper with visible grains ________ if yes, you have sandstone, go to 9. If no, go to 8.
8. Does rock have clearly visible particles, either rounded or angular? ________ You have either conglomerate or breccia, check with reference table, then go to 9.
9. The rock is ____________________________

Rock #4

1. Is the rock dark in color, and feels like wood? ________ If yes, you have coal, go to 9. If no, go to 2.
2. Does the rock bubble with acid? _______ If yes, you have limestone, go to 9. If no, go to 3.
3. Is the rock clear with a definite cube like shape? ________ If yes, you have rock salt, go to 9. If no, go to 4.
4. Can you scratch the rock with your fingernail? ________ If yes, you have rock gypsum, go to 9. If no, go to 5.
5. Is the rock very smooth with no visible grains? ________ If yes, you have shale, go to 9. If no, go to 6.
6. Does the rock have a gritty feel but the grains are difficult to see? ________ If yes, you have siltstone, go to 9. If no, go to 7.
7. Does the rock have a texture of sandpaper with visible grains ________ if yes, you have sandstone, go to 9. If no, go to 8.
8. Does rock have clearly visible particles, either rounded or angular? _______ You have either conglomerate or breccia, check with reference table, then go to 9.
9. The rock is ____________________________

Rock #5

1. Is the rock dark in color, and feels like wood? ________ If yes, you have coal, go to 9. If no, go to 2.
2. Does the rock bubble with acid? _______ If yes, you have limestone, go to 9. If no, go to 3.
3. Is the rock clear with a definite cube like shape? ________ If yes, you have rock salt, go to 9. If no, go to 4.
4. Can you scratch the rock with your fingernail? ________ If yes, you have rock gypsum, go to 9. If no, go to 5.
5. Is the rock very smooth with no visible grains? ________ If yes, you have shale, go to 9. If no, go to 6.
6. Does the rock have a gritty feel but the grains are difficult to see? ________ If yes, you have siltstone, go to 9. If no, go to 7.
7. Does the rock have a texture of sandpaper with visible grains ________ if yes, you have sandstone, go to 9. If no, go to 8.
8. Does rock have clearly visible particles, either rounded or angular? _______ You have either conglomerate or breccia, check with reference table, then go to 9.
9. The rock is ____________________________

Observations:

1. Name one characteristic that clastic sedimentary rocks have in common.

2. From your answer in question 1, describe what the characteristic is and how it helps in identifying the rocks.

3. What is the fundamental difference between siltstone and shale?

4. What is the fundamental difference between conglomerate and breccia?

5. What type of igneous rock do these clastic sedimentary rocks appear to be formed from?

Conclusions:

6. What is a unique feature of crystalline evaporate rock?

7. What portion of the exposed rock on the Earth's surface is sedimentary rock?

8. Since we know what limestone is made of, why would an area rich in limestone be rich in caves or caverns? (Hint: Look at the composition)

9. Most fossils are found in sedimentary rock. What is it about the origin of sedimentary rock that makes this so?

10. You want to build a house on a piece of lake front property that you own. On your property you have a hillside with sandstone at one elevation. Higher up on the slope the layers are shale. Lower levels are closer to the lake. If you build a house near the top of the hill you have a view of the lake and can see beautiful sunrises and sunsets. Which elevation would you choose and why?

11 Metamorphic Rock

NAME ______________________

Metamorphic rock is a rock that has changed because it was exposed to intense heat, pressure or both. Metamorphic rock changes because the crystals that make up the rock change and get distorted. We know that the internal atomic structure of a mineral is responsible for its properties, so any changes actually change the mineral itself. Metamorphic rocks are classified into two groups, foliated and non-foliated. Foliated refers to minerals that align themselves in the rock in what is described as flat. It gives the appearance of how autumn leaves pile flat on top of each other when they fall. The term actually comes from leaves; foliage. Nonfoliated rocks have their minerals distorted and changed and yet the minerals do not align themselves.

Before Metamorphism

After Metamorphism

Objective:
The objective of this lab is to identify the five different metamorphic rocks given in the sample.

Essential Question:

____

____

Materials:
- Sample kit of 5 rocks
- Scheme for Metamorphic Rock Identification chart on page 48
- Magnifying lens

Procedure:

With the assistance of a magnifying glass, identify the rocks using the Scheme for Metamorphic Rock Identification chart and by following the flow chart.

Scheme for Metamorphic Rock Identification

TEXTURE		GRAIN SIZE	COMPOSITION	TYPE OF METAMORPHISM	COMMENTS	ROCK NAME	MAP SYMBOL
FOLIATED	MINERAL ALIGNMENT	Fine	MICA	Regional (Heat and pressure increases) ↓	Low-grade metamorphism of shale	**Slate**	
FOLIATED	MINERAL ALIGNMENT	Fine to medium	MICA, QUARTZ, FELDSPAR, AMPHIBOLE, GARNET	Regional (Heat and pressure increases) ↓	Foliation surfaces shiny from microscopic mica crystals	**Phyllite**	
FOLIATED	MINERAL ALIGNMENT	Fine to medium	MICA, QUARTZ, FELDSPAR, AMPHIBOLE, GARNET, PYROXENE	Regional (Heat and pressure increases) ↓	Platy mica crystals visible from metamorphism of clay or feldspars	**Schist**	
FOLIATED	BAND-ING	Medium to coarse	MICA, QUARTZ, FELDSPAR, AMPHIBOLE, GARNET, PYROXENE	Regional (Heat and pressure increases) ↓	High-grade metamorphism; mineral types segregated into bands	**Gneiss**	
NONFOLIATED		Fine	Carbon	Regional	Metamorphism of bituminous coal	**Anthracite coal**	
NONFOLIATED		Fine	Various minerals	Contact (heat)	Various rocks changed by heat from nearby magma/lava	**Hornfels**	
NONFOLIATED		Fine to coarse	Quartz	Regional or contact	Metamorphism of quartz sandstone	**Quartzite**	
NONFOLIATED		Fine to coarse	Calcite and/or dolomite	Regional or contact	Metamorphism of limestone or dolostone	**Marble**	
NONFOLIATED		Coarse	Various minerals	Regional or contact	Pebbles may be distorted or stretched	**Metaconglomerate**	

Rock #1

1. Is the rock foliated? ______________________________
   If yes, continue to question 2. If no, skip to question 3.

2. What is the grain size? (Circle one) Fine Fine to Medium Medium to Course Coarse
   Continue to number 5.

3. Is the rock texture coarse with visible pebbles? ______________________
   If yes, skip to number 5. If no, continue to question 4.

4. Is the rock fine to coarse grained? __________ Continue to number 5.

5. The rock is______________________________

6. Comments about the rock: ______________________________

____

____

Rock # 2

1. Is the rock foliated? ______________________
   If yes, continue to question 2. If no, skip to question 3.

2. What is the grain size? (Circle one) Fine Fine to Medium Medium to Course Coarse
   Continue to number 5.

3. Is the rock texture coarse with visible pebbles? ______________________
   If yes, skip to number 5. If no, continue to question 4.

4. Is the rock fine to coarse grained? __________ Continue to number 5.

5. The rock is______________________

6. Comments about the rock: ______________________

   ____

   ____

Rock # 3

1. Is the rock foliated? ______________________
   If yes, continue to question 2. If no, skip to question 3.

2. What is the grain size? (Circle one) Fine Fine to Medium Medium to Course Coarse
   Continue to number 5.

3. Is the rock texture coarse with visible pebbles? ______________________
   If yes, skip to number 5. If no, continue to question 4.

4. Is the rock fine to coarse grained? __________ Continue to number 5.

5. The rock is______________________

6. Comments about the rock: ______________________

   ____

   ____

Rock #4

1. Is the rock foliated? ______________________________
   If yes, continue to question 2. If no, skip to question 3.
2. What is the grain size? (Circle one) Fine Fine to Medium Medium to Course Coarse
   Continue to number 5.
3. Is the rock texture coarse with visible pebbles? ______________________
   If yes, skip to number 5. If no, continue to question 4.
4. Is the rock fine to coarse grained? __________ Continue to number 5.
5. The rock is______________________________
6. Comments about the rock: ______________________________

____

____

Rock #5

1. Is the rock foliated? ______________________________
   If yes, continue to question 2. If no, skip to question 3.
2. What is the grain size? (Circle one) Fine Fine to Medium Medium to Course Coarse

   Continue to number 5.
3. Is the rock texture coarse with visible pebbles? ______________________
   If yes, skip to number 5. If no, continue to question 4.
4. Is the rock fine to coarse grained? __________ Continue to number 5.
5. The rock is______________________________
6. Comments about the rock: ______________________________

____

____

Observations:

1. What are the two main types of metamorphic rock in relation to its structure?

2. What does foliation mean?

3. Metamorphic rocks are transformed by heat and pressure from what rock types?

4. What type of metamorphic rock will you get from limestone?

5. What type of metamorphism do you get over a large area?

6. What type of rock will have discreet sections of minerals in coarse grains?

7. What type of rock has undergone the most intense heat and pressure?

Look at the chart below. It indicates the depth at which metamorphic rock forms. Notice that sedimentary rock is at the surface. The material that is shaded is rock that has melted. You will need to use this chart to answer questions 8–10 on this page.

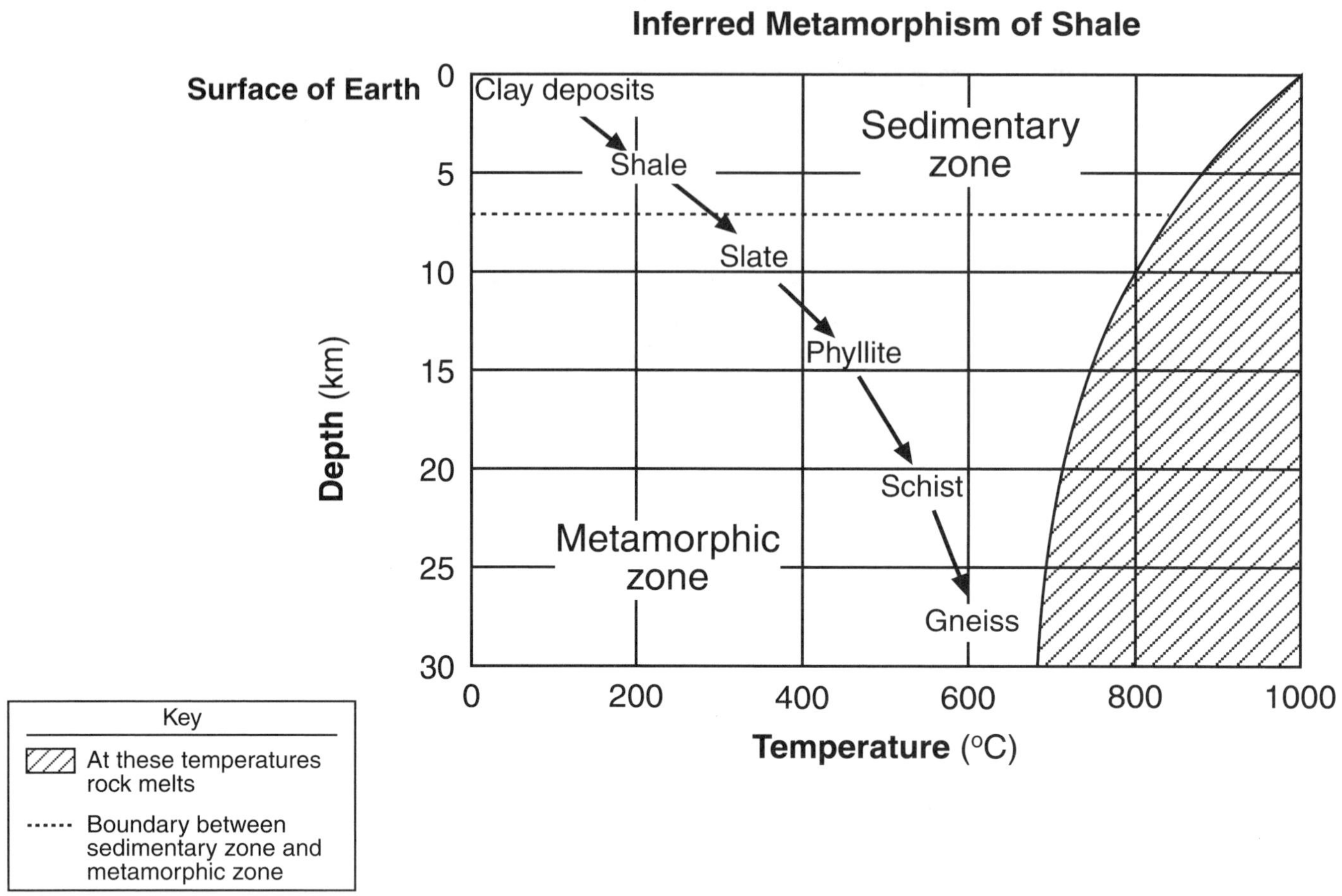

8. What is the difference between schist and gneiss?

____

____

9. What type of rock is formed exclusively from contact metamorphism?

____

____

10. What type of rock is extremely flaky and breaks easily?

____

____

12 Earthquake Location

An earthquake is any shaking on the Earth's surface. There are many different causes for the Earth's surface to shake. Many times we can't even feel the earthquakes because they are so slight. These are only detectable through sensing devices called seismographs. Sometimes the Earth's surface can shake from man-made events. A seismograph can record traffic, road construction, explosions and even the collapse of buildings.

Plate motion is the most common cause of earthquakes. Earthquakes occur along plate boundaries. New York State sits in the middle of the North American plate and yet the northern border of New York State is highly prone to large earthquakes.

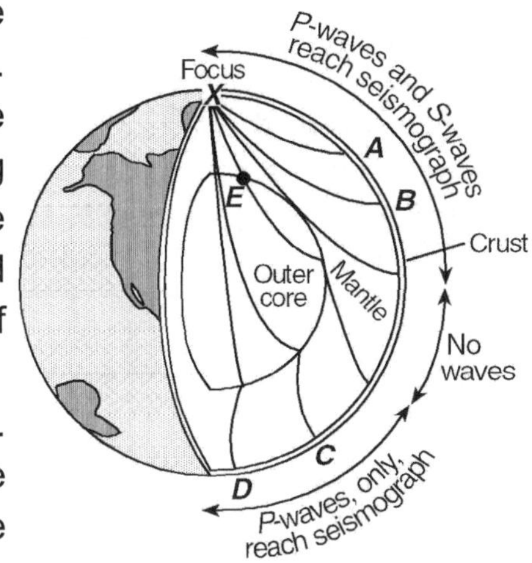

Objective:

The objective of this investigation is to understand how earthquakes are located.

Essential Question:

____

____

Hypothesis:

If____

Then____

Because____

Variables:

Independent:____

Dependent:____

Procedure: Use the Earthquake P-Wave and S-Wave Travel time chart below to answer the following question.

Directions: Using the Earthquake P-wave and S-wave Time chart below, calculate the distance to the epicenter from Palisades, NY and Newport, VA. Place your answers in the spaces provided in Report Sheet #1.

Report Sheet # 1

Station Time	P-Wave Travel Time	Distance to Epicenter	S-Wave Travel Time	Distance to Epicenter
Palisades, NY,	00:03:20	km	00:06:00	km
Newport, VA	00:07:50	km	00:14:00	km

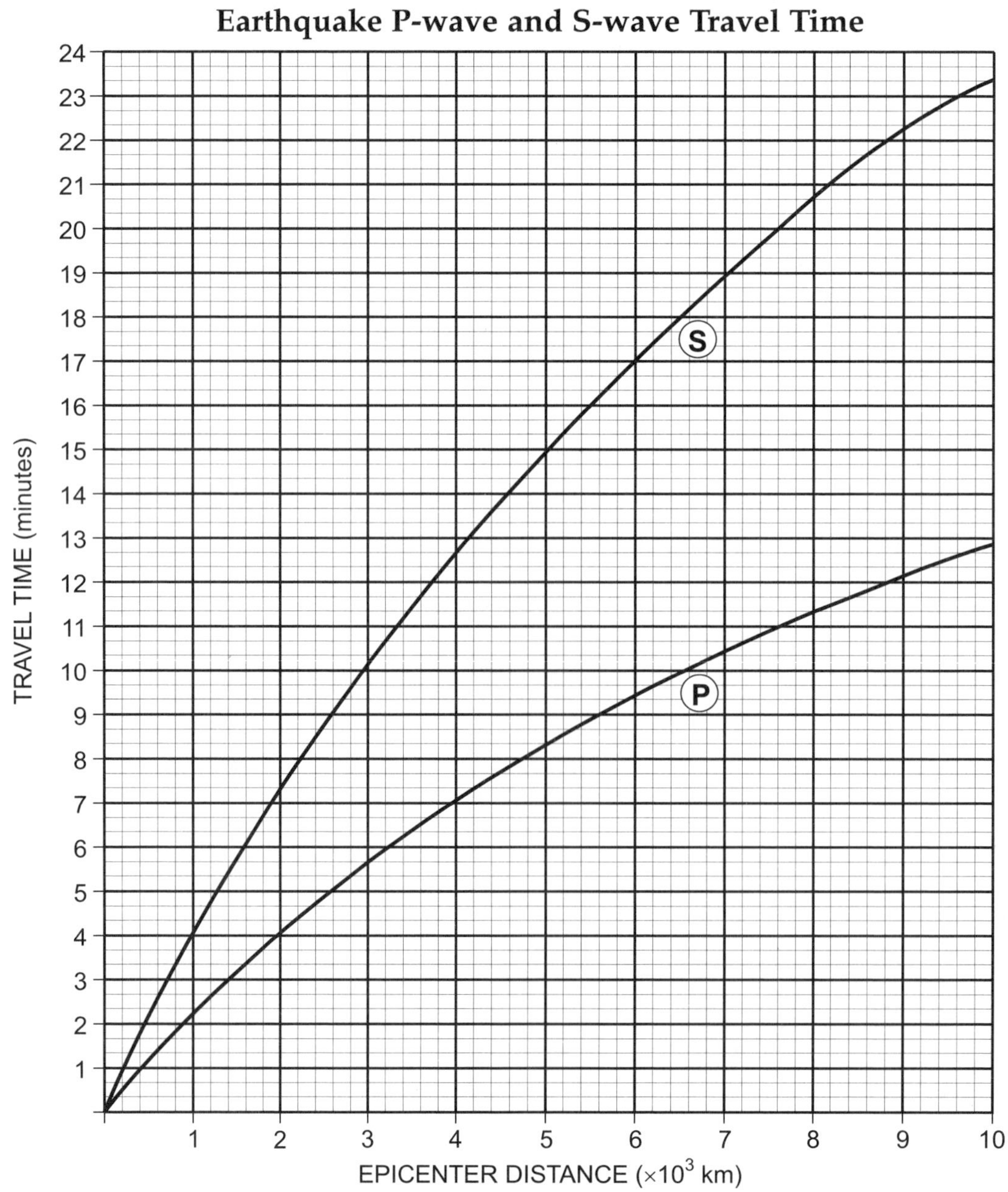

Directions: In the chart below, fill in the information from the seismographs. Use this information to plot the epicenter on the map on page 56. You will need to refer to the Earthquake P-wave and S-wave Travel Time chart on page 54

NOTE: The time is listed as hours : minutes : seconds. Reminder: there is only 60 minutes in an hour and 60 seconds in a minute. This is immensely important when adding or subtracting. Fill in the blanks accordingly.

Report Sheet # 2

Seismic Station	"P" Arrival Time	"S" Arrival Time	S" Time minus "P" Time	"P" Wave Travel Time	Distance to Epicenter
Station X Annapolis, MD	: :	: :	: :	: :	km
Station Y Oklahoma City, OK	: :	: :	: :	: :	km
Station Z Des Moines, IA	: :	: :	: :	: :	km

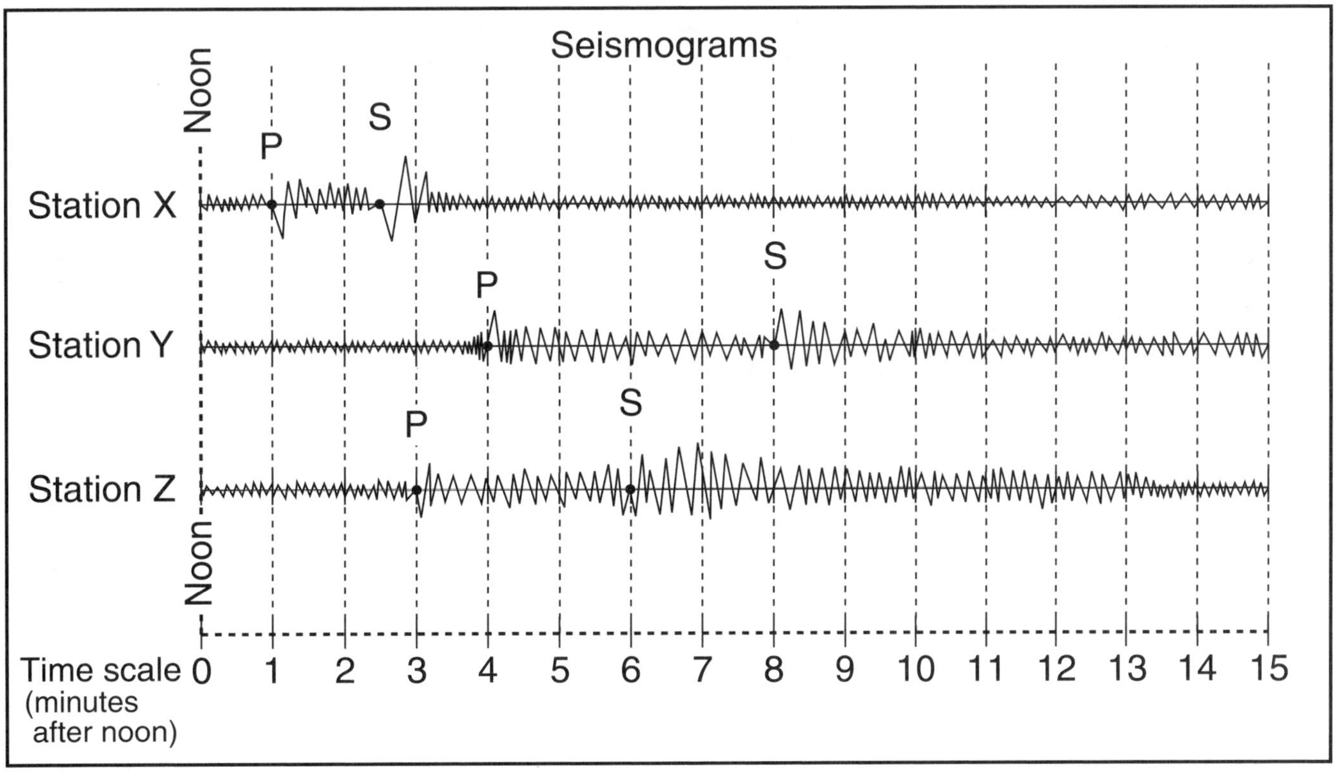

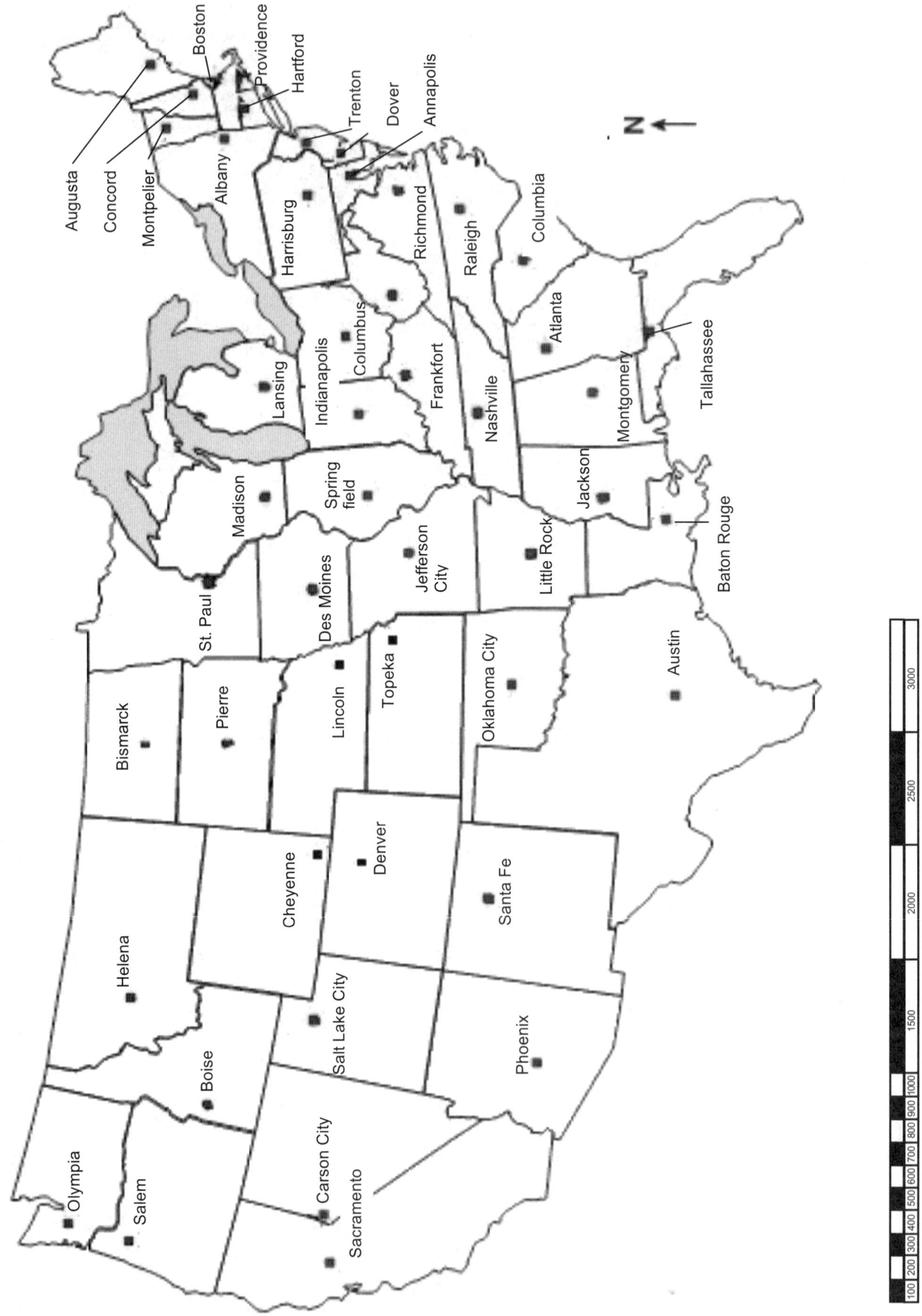
N
Augusta
Concord
Montpelier
Boston
Providence
Hartford
Albany
Trenton
Dover
Annapolis
Harrisburg
Richmond
Raleigh
Columbia
Columbus
Lansing
Indianapolis
Frankfort
Nashville
Atlanta
Tallahassee
Montgomery
Madison
Spring field
Jackson
Baton Rouge
St. Paul
Des Moines
Jefferson City
Little Rock
Bismarck
Pierre
Lincoln
Topeka
Oklahoma City
Austin
Helena
Cheyenne
Denver
Santa Fe
Salt Lake City
Boise
Phoenix
Olympia
Salem
Carson City
Sacramento
100
200
300
400
500
600
700
800
900
1000
1500
2000
2500
3000

Observations:

1. What state is the epicenter located in?

2. Determine the city closest to the epicenter by using the Earth Science Reference Tables in the back of the book.

3. What information would only one seismic station give?

4. What information would only two seismic stations give?

5. If you did not use a drawing compass, indicate how you would know which station was closest to the earthquake.

6. Why does the time between the arrival of the P-wave and the S-wave become greater and greater as you get farther away from the epicenter?

7. There are seismic stations all over the world. Explain how these stations can record data on the other side of the planet.

Directions: Look at the following illustration. Answer the following questions.

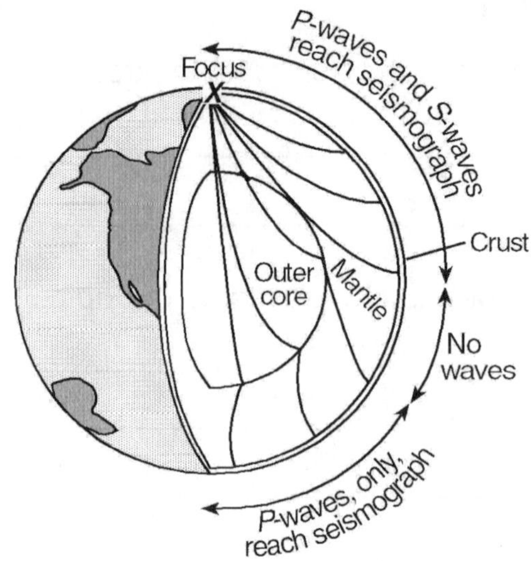

8. Explain why seismic stations don't always record data from every earthquake.

____

____

____

9. What other events do seismograms record?

____

____

10. What other information do scientists look to seismic waves for?

____

____

____

____

NAME ____________________

13 Volcano Location

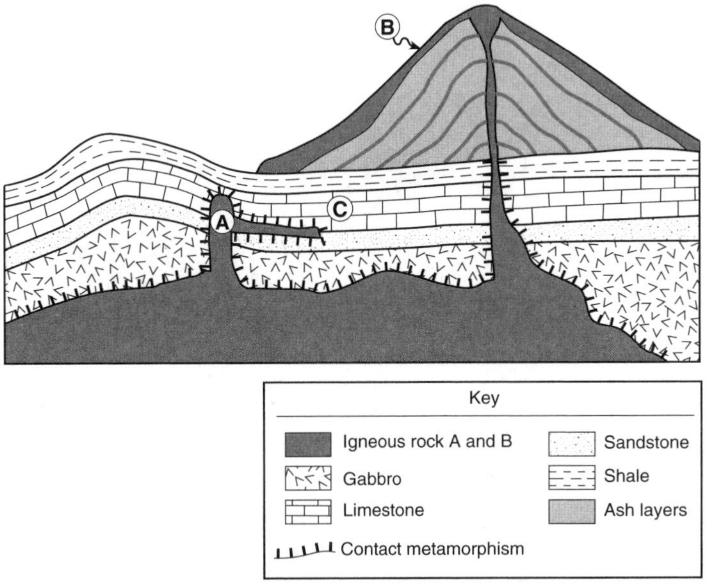

A volcano is an area where molten magma breaks through the surface. When magma breaks the surface of the Earth it is called lava. Lava tends to build up and form a cone by cooling and forming additional layers. Most of the volcanoes on Earth are underwater and in areas known as mid-oceanic ridges.

Volcanoes are classified by their composition and shape. A cinder cone volcano is the smallest and is made of loose debris, blown out during an eruption. A composite volcano is made from alternating layers of debris, cinders, and hardened lava. Composite volcanoes are the most common type of active volcanoes on Earth and make up more than 75 percent of the volcanoes that have erupted in recorded history. The largest volcano is a shield volcano. The shield volcano is associated with basaltic volcanoes and is known to be gentler than the other two, but have had violent eruptions. Mauna Kea of Hawaii is a shield volcano and is considered the largest mountain on the planet. It is considered so because the base of the mountain is at the bottom of the Pacific Ocean.

Objective:

The objective of this lab is to find if a pattern exists to the world's volcanoes.

Materials:

- Volcanoes and Their Locations chart on page 61
- Red marker or pencil

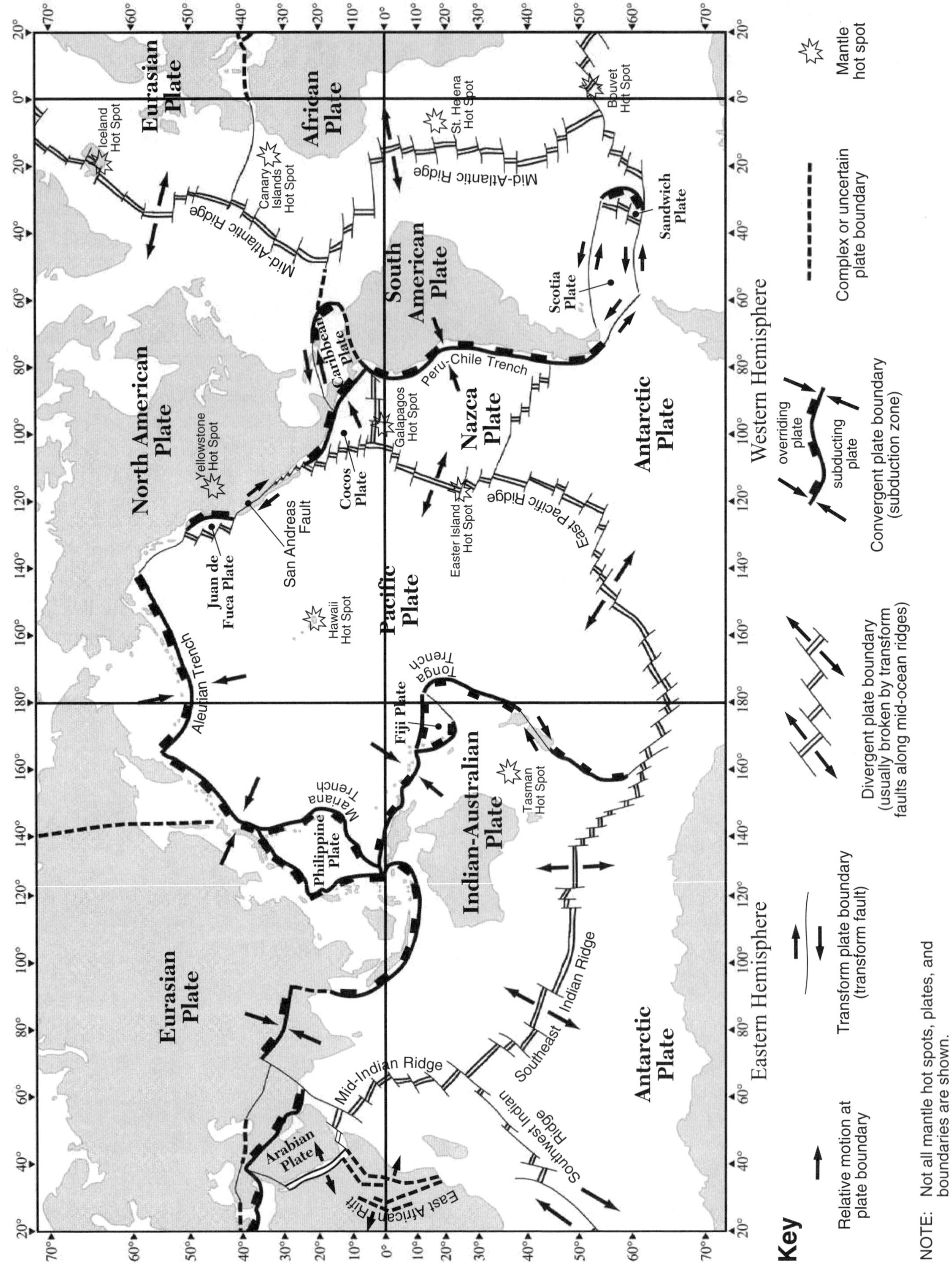

Eurasian Plate
African Plate
North American Plate
South American Plate
Caribbean Plate
Cocos Plate
Nazca Plate
Pacific Plate
Juan de Fuca Plate
Philippine Plate
Fiji Plate
Indian-Australian Plate
Arabian Plate
Scotia Plate
Sandwich Plate
Antarctic Plate
Iceland Hot Spot
Canary Islands Hot Spot
St. Helena Hot Spot
Bouvet Hot Spot
Yellowstone Hot Spot
Galapagos Hot Spot
Easter Island Hot Spot
Hawaii Hot Spot
Tasman Hot Spot
Mid-Atlantic Ridge
Peru-Chile Trench
East Pacific Ridge
San Andreas Fault
Aleutian Trench
Tonga Trench
Mariana Trench
Mid-Indian Ridge
Southeast Indian Ridge
Southwest Indian Ridge
East African Rift
Eastern Hemisphere
Western Hemisphere
Key
Relative motion at plate boundary
Transform plate boundary (transform fault)
Divergent plate boundary (usually broken by transform faults along mid-ocean ridges)
Convergent plate boundary (subduction zone)
overriding plate
subducting plate
Complex or uncertain plate boundary
Mantle hot spot
NOTE: Not all mantle hot spots, plates, and boundaries are shown.

Procedure:

Mark the following volcanoes on the map on page 60 by making a small triangle in the appropriate place. Reminder: It is still called "east" to the right of the Prime Meridian on the map.

Volcanoes and Their Locations

1. Yasur	19°S	169°E	2. Merapi	7°S	110°E
3. Kerinci	1°S	101°E	4. Langila	5°S	148°E
5. Ol Doinyo	2°S	36°E	6. South Sister	44°N	121°W
7. Sakura-Jima	31°N	130°E	8. Kavachi Seamount	9°S	158°E
9. Pinatubo	15°N	120°E	10. Redoubt	60°N	152°W
11. Nyamuragria	1°N	124°E	12. Hachijo-Jima	33°N	139°E
13. Veniaminof	56°N	159°W	14. Mauna Loa	19°N	155°W
15. Lokon Sulawesi	1°N	124°E	16. Guagua Pichincha	1°S	78°W
17. Krakatau	6°S	105°E	18. Llaima	38°S	71°W
19. Barren Island	12°N	94°E	20. Manam	4°S	145°E
21. Kick-'em Jenny	12°N	61°W	22. San Cristobal	12°N	87°W
23. Pago	5°S	150°E	24. Arenal	10°N	84°W
25. Papandayan	7°S	107°E	26. Tangkubanparahu	6°S	107°E
27. Erta Ale	13°N	40°E	28. Villarrica	39°S	71°W
29. Bezymianny	56°N	160°E	30. Raung	2°N	125°E

Observations:

1. What is a volcano?

____

____

2. What is the most common type of volcano?

____

____

3. What is the most common structure of volcanoes?

4. What are the three types of volcanoes?

5. Is there a pattern to the volcano distribution around the planet?

6. Do the volcanoes' pattern coincide with other geologic events?

7. What type of volcano is associated with convergent plate boundaries?

8. Describe what type of plate boundary is most closely associated with most volcanoes.

9. Are volcanoes likely to form in your area?

10. Explain why earthquakes and volcanoes appear in generalized belts around the planet.

NAME ______________________

14 Pangea

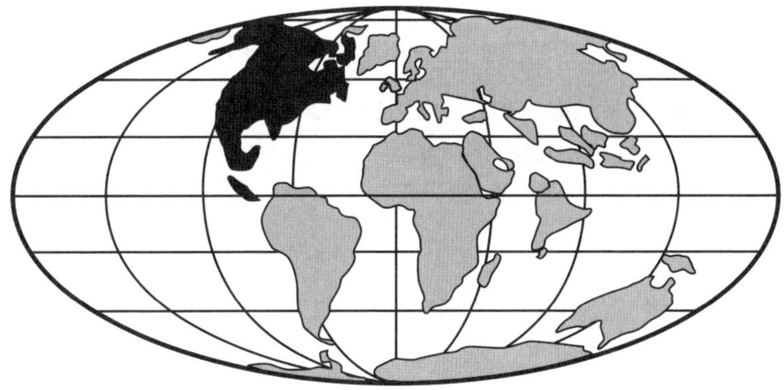

Pangea is the name given to the giant super continent that existed hundreds of millions of years ago. Alfred Wegner first proposed the concept of the super continent. He was looking closely at the east coast of South America and the west coast of Africa. He noticed that the two continents seem to fit like a giant jigsaw puzzle. He also cited fossil evidence. Scientists found the fossils of the mesosaurus, an ancestor of the alligator in both South America and Africa. Yet, the mesosaurus did not swim. Additional fossils have since been found on both continents. Other evidence for Pangea was also found on continents separated by oceans, such as terminations of mountain ranges and similar rock types. Later, in the 1940's SONAR was developed. SONAR was later used to discover the mid-oceanic ridges which proved the separation of Pangea. In this investigation, we will be going backward, not forward to reconstruct Pangea.

Objective:

The objective of this investigation is to reconstruct Pangea using geologic and fossil evidence.

Essential Question:

____

____

Materials:

- Scissors
- Glue
- Pangea Cutouts on page 69
- Inferred Positions of the Earth's Landmasses chart in back of book

Procedure:

1. Tear out the Pangea Cutouts on page 69.
2. Cut out the continents.
3. Use the Inferred Positions of the Earth's Landmasses chart found on page 211.
4. On page 68, arrange the continents in a way that would form the continent Pangea and glue them down.

Observations:

1. What was on the paper that helped to make the continents line up?

____

____

2. What are some of the ways in which you were able to position the continents?

____

____

3. Did the continents fit perfectly?

____

____

4. How long ago did Pangea exist?

____

____

5. What are some ways in which the idea of Pangea first came about?

____

____

6. Describe the difficulties with positioning the current continents into Pangea.

____

____

____

____

7. Look at the Tectonic Plates map from the Earth Science Reference Tables in the back of the lab workbook. Based on the map, which continent is currently spreading into two?

____

8. Describe the direction in which North America is moving in relation to Asia.

____

____

____

____

9. Describe the direction in which North America is moving in relation to Europe.

____

____

____

____

10. Knowing that the continents are moving, choose which one of the following models would most likely be Earth's position in a million years.

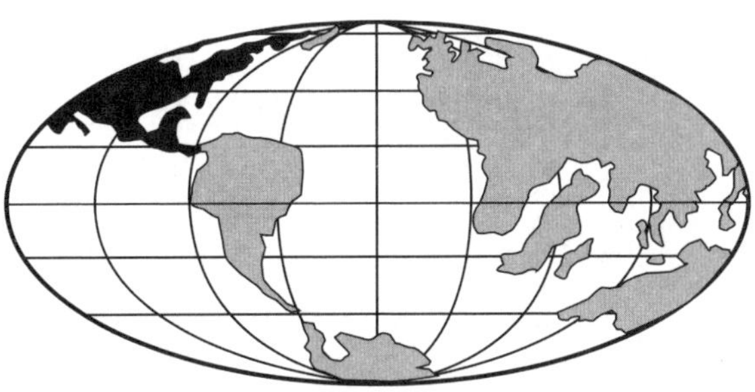

(1)

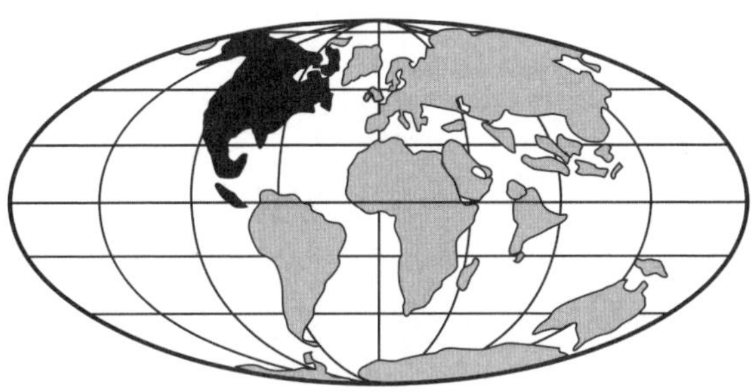

(3)

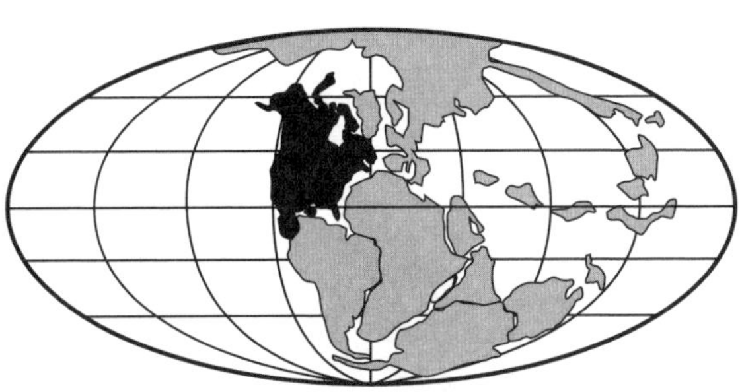

(2)

(4)

10__________

Pangea Cutouts paste here

Pangea Cutouts

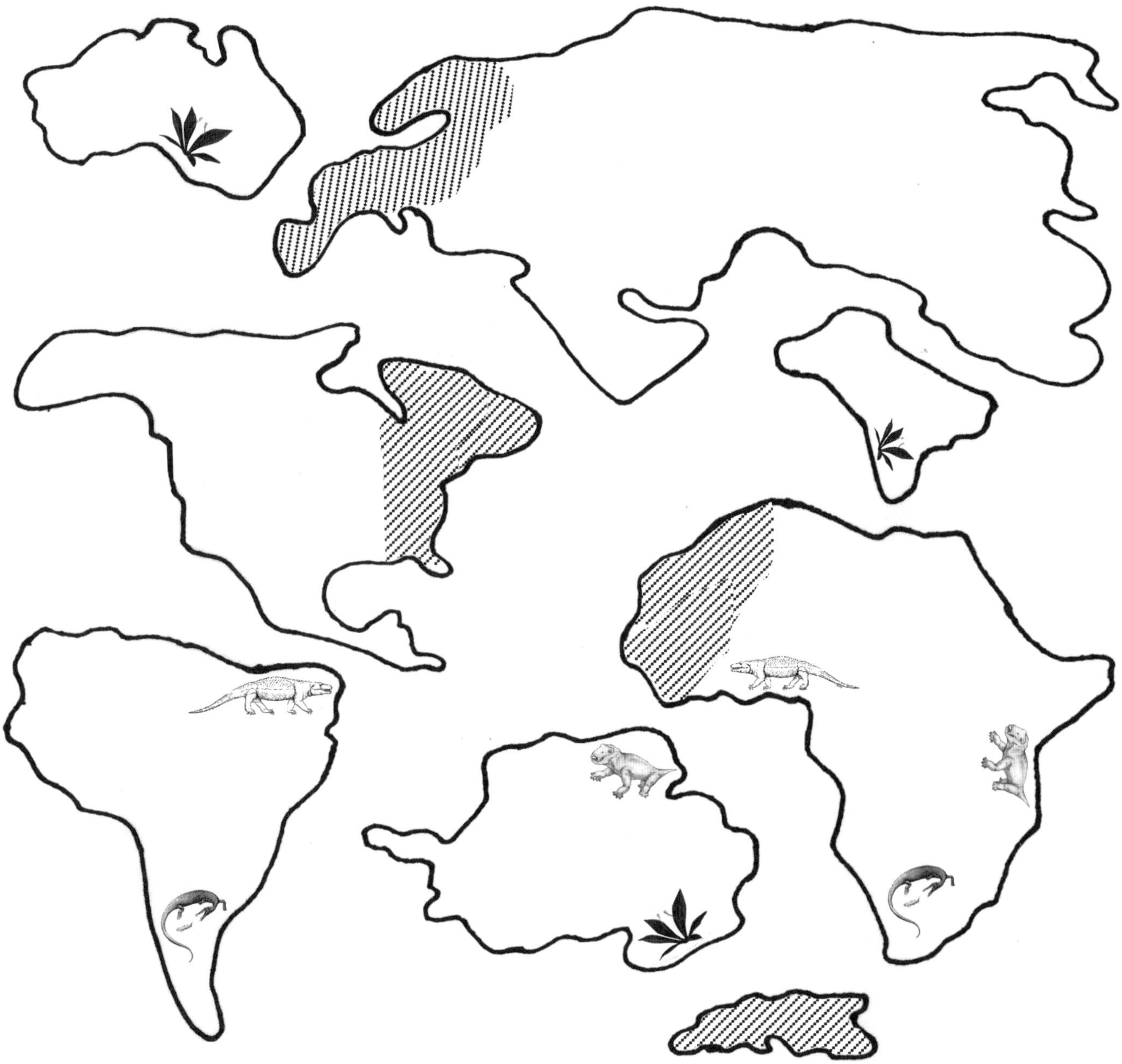

NAME ______________________________

15 Chemical Weathering

Chemical weathering takes place when rocks break apart as a result of chemical changes. Oxidation is the chemical reaction of oxygen with other substances. Most weathering is caused by hydrolysis, which is water reacting with other materials. Carbonation occurs when carbonic acid reacts with the minerals. Plants make acids which cause chemical weathering. The rate at which chemical weathering occurs varies largely in part to the particle size of the material being broken down. Climate also determines the speed at which parent materials undergo chemical weathering. Chemical weathering is fastest in hot, moist climates and slowest in cold, dry climates.

****CAUTION! THIS LAB EMPLOYS THE USE OF AN ACID, USE EXTREME CAUTION.****

Objective:

The objective of this lab is for students to infer information regarding chemical weathering based on observations.

Essential Question:

____

____

Hypothesis:

If ____

Then ____

Because ____

Variables:

Independent: ____

Dependent: ____

Materials:

- Beakers
- Acid
- Chalk
- Index card
- Hammer
- Eye protection

Procedure:

1. Label one beaker *A* and the other *B*.
2. Fill each beaker with 2.5 cm of acid.
3. Break chalk in half and set one-half aside. Take the other half and crush it on the index card.
4. Simultaneously place a whole piece of chalk in Beaker *A* and the crushed particles in Beaker *B*.
5. Record time.
6. Make observations in intervals of thirty seconds. Record findings.

Time	Beaker A	Beaker B
00:00		
00:30		
01:00		
01:30		
02:00		
02:30		
03:00		
03:30		
04:00		
04:30		
05:00		

Observations:

1. Describe what happened to the material in the Beaker *A*.

2. Describe what happened to the material in the Beaker *B*.

3. What were the differences in the weathering of the two beakers?

4. What were the similarities in the weathering of the two beakers?

5. Which beaker's particles reacted more quickly?

6. Describe the differences in reaction time between Beaker *A* and Beaker *B*?

7. What can you infer about chemical weathering and particle size?

8. Extrapolate: What type of rock is similar in composition to chalk?

9. What type of weathering would you associate with this type of rock?

10. Describe how a rock can be weathered by more than one process.

NAME ______________________

16 Physical Weathering

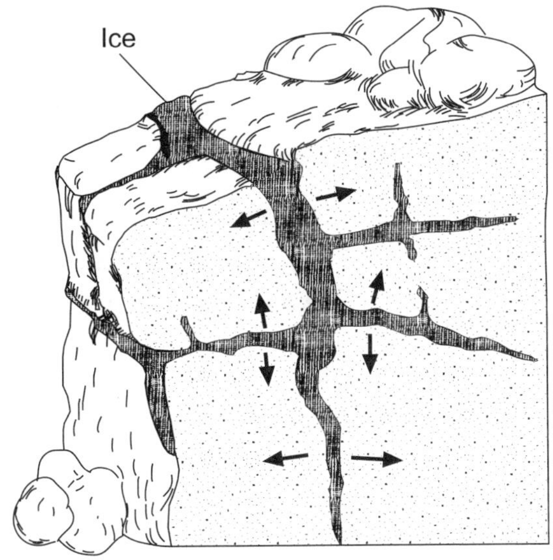

Climate directly affects soil formation. Water, ice, wind, heat, and cold can cause physical weathering. Physical weathering is caused by the loosening and breaking up of rocks. Rocks are worn down by water and wind and ground to bits by the slow movement of glaciers. Water in rock crevices expands when it freezes, causing the rocks to crack. It is through this expansion of water that wedges rocks in two. This kind of physical weathering is often called frost wedging.

Objective:

The objective of this lab is for students to infer information regarding physical weathering based on their observation.

Essential Question:

____

____

Hypothesis:

If ____

Then ____

Because ____

Variables:

Independent: ____

Dependent: ____

Materials:
- Beaker (250 mL)
- Food coloring
- Halite
- Test tube
- Crushed ice
- Marker
- Ruler

Procedure:

1. Place 100 mL of crushed ice in beaker. Cover ice with halite. Set aside.
2. Pour 10 mL of colored water in test tube. Mark water level on test tube with marker.
3. Put test tube in beaker, stir ice with the test tube. Continue to do so until you see a change occur inside the test tube. Use marker to record your findings.
4. Record differences in water level before and after.
5. In the boxes below, draw a before and after picture of the test tube.

DRAW A BEFORE AND AFTER PICTURE OF THE TEST TUBE	
Before	**After**

Observations:

1. What happened to the water inside the test tube?

____

____

____

____

____

____

2. What changes occurred after the experiment was completed?

____

____

3. What is the difference from the beginning mark and the final “water level”? (Hint: subtraction)

____

4. Describe what the water in the test tube looks like.

____

____

5. What do you think would happen to rocks if water seeped into cracks and froze?

____

____

____

____

____

____

____

6. What types of climates would have this type of weathering?

____

____

7. Would this weathering be a problem in a dry region? Explain.

____

____

____

____

____

____

____

8. What is a common problem where we live that is associated with this type of weathering?

9. Extrapolate: How does this experiment illustrate mechanical weathering?

10. What do people do to prevent this type of weathering?

NAME ____________________

17 Rock Abrasion

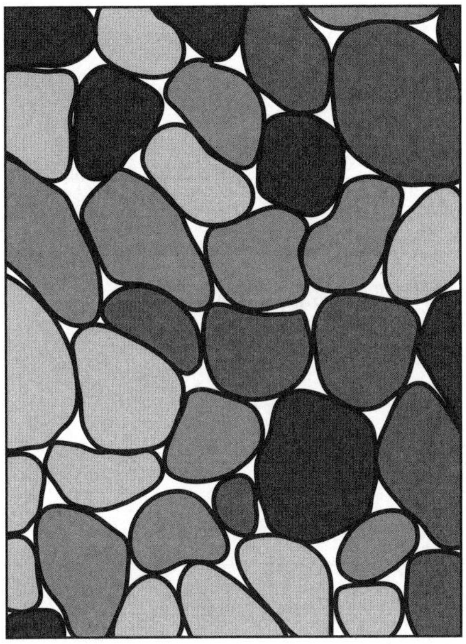

River stones change shape by the tumbling and abrasion in a riverbed and the transportation of sediments in the river. River stones are rounded rocks that are weathered by agents of erosion. River stones are found along streambeds and are extremely smooth. The rocks become increasingly rounded as they are exposed to agents of weathering.

Different kinds of weathering can create different shapes to the rocks. A rock weathered from frost action results in angular fragments. These rocks are wedged apart because of the freeze thaw cycle. These rocks tend to be sharp and irregularly shaped.

Objective:

The objective of this lab is to determine how objects change in shape due to weathering.

Essential Question:

____

____

Hypothesis:

If____

Then____

Because____

Variables:

Independent:____

Dependent:____

Materials:

- Container
- Stopwatch
- 10 sugar cubes
- Ruler

Procedure:

1. In the top box of the chart, draw a picture of the sugar cubes.
2. Describe the cubes. (size, shape, etc)
3. Place the sugar cubes into the container.
4. Shake the container vigorously for 30 seconds.
5. Draw and describe what you see in the spaces provided in the chart.
6. Repeat steps 4 and 5 until all 10 observations are completed.

Drawing	Description
Initial Observation 0:00	
1st Observation 0:30	
2nd Observation 1:00	
3rd Observation 1:30	
4th Observation 2:00	

5th Observation 2:30	
6th Observation 3:00	
7th Observation 3:30	
8th Observation 4:00	
9th Observation 4:30	
10th Observation 5:00	

Observations:

1. Describe the overall shape of the objects at the beginning of the exercise.

2. Describe the overall shape of the objects at the end of the exercise.

3. Approximately how long did it take the objects to change shape?

4. What caused the objects to change shape?

5. Explain how this lab activity relates to weathering of river stones.

6. As the objects were shaken, think about how the container was shaken. What do you think the outcome would be if the container were shaken less vigorously?

7. Using your answer to question 6, explain how you reached that conclusion.

____

____

____

____

8. What do you think the outcome would be if the container were shaken more vigorously?

____

____

9. Using your answer to question 8, explain how you reached that conclusion.

____

____

____

____

10. Please explain how river stones are formed. Fully explain your answer.

____

____

____

____

____

____

____

____

____

____

____

NAME ______________________

18 Settling Rate

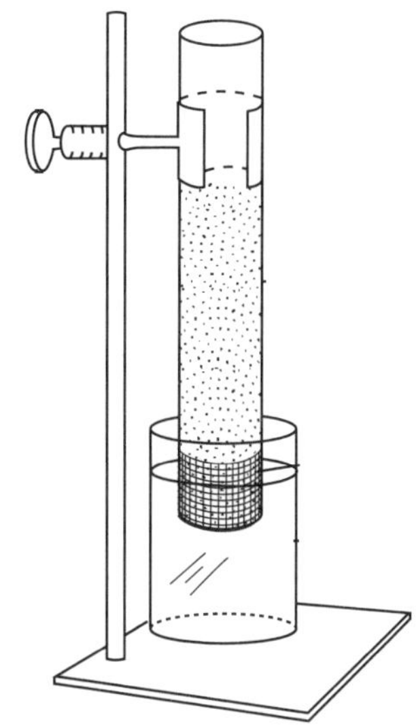

The process by which sediments are released, settled from, or dropped from an erosional system is called deposition. Deposition is part of a series of processes that start with the formation and movement of sediments by weathering and erosion. In deposition, the sediments are placed in locations where they may form beds, or layers, of sedimentary rocks. Most final deposition occurs in large bodies of water. Deposition usually occurs when the velocity, or speed, of the stream, wind, or other erosional system decreases or just stops moving. The time it takes for sediments to be deposited, the rate of deposition, varies. The faster the rate of deposition, the less time deposition takes; and the slower the rate of deposition, the more time deposition takes. Two types of factors, the velocity of the system and characteristics of the sediments themselves, affect the rate of deposition.

Objective:

The objective of this lab is to determine how objects of equal density and shape will settle in a liquid.

Essential Question:

____

____

Hypothesis:

If ____

Then ____

Because ____

Variables:

Independent: ____

Dependent: ____

Materials:

- Graduated cylinder
- Stop watch
- Clear liquid laundry detergent
- White glass beads; varying sizes

Procedure:

1. Mark off a 100 ml section of the graduated cylinder.
2. Fill the graduated cylinder with clear liquid laundry detergent to the 100 ml mark.
3. Drop in one of the beads.
4. Time the bead as it passes from the 100 ml line that you've marked to the bottom.
5. Record your findings in the chart below.
6. Repeat steps 3 – 5 until completed.

	Large Bead	Medium Bead	Small Bead
1st Trial	:	:	:
2nd Trial	:	:	:
Total	:	:	:
Average	:	:	:

Observations:

1. Which size bead dropped the fastest?

____

____

2. Which bead size had the greatest variance?

____

____

3. What is the pattern, if any?

____

____

4. When thinking of rock sediment, what size particle will settle faster in a body of water?

____

____

5. What can you infer about deposition?

____

____

For questions 6 through 9, answer the questions and place your answer in the space provided. Then draw a simple graph representing the relationship in space provided.

6. As an object becomes more angular, the settling rate

   a) increases
   b) decreases
   c) stays the same

6________

7. As the density of an object increases, the settling rate

   a) increases
   b) decreases
   c) stays the same

7________

8. As the size of an object increases, the settling rate

   a) increases
   b) decreases
   c) stays the same

8 ________

9. As the size of the glass beads increased, what happened to the density?

   a) increases
   b) decreases
   c) stays the same

9 ________

10. Explain how the size of different materials affect this Settling Rate lab.

NAME ____________________

19 Landscapes

A landscape is a region on Earth's surface. The shape and makeup of the landscape is formed by the climate, local bedrock, and human activities. The relief of landscapes is described as mountains, plateaus and plains. In New York State, the landscape is not only determined by geology and climate but has also been deeply affected by the glaciers. When we look at the landscape regions of New York State, we must look to see the underlying geology. Mountainous regions have metamorphosed bedrock in New York State. What appear to be mountains may not, in fact, be mountains and areas that appear to be small foothills are the remains of once tall mountains. To truly determine the landscape region, we must look underneath the surface.

Objective:

The objective of this lab is to apply information on landscapes of New York State and to understand the topography of the state.

Essential Question:

____

____

Materials: • Colored pencils

Procedure:

1. In the following inquiry, compare the bedrock of various areas of New York State.
2. Compare the landscape regions based on the underlying bedrock.
3. On the Landscape Regions map on page 91, color the mountain regions red, the plateau regions green and the plains regions blue.

Generalized Bedrock Geology of New York State

modified from
GEOLOGICAL SURVEY
NEW YORK STATE MUSEUM
1989

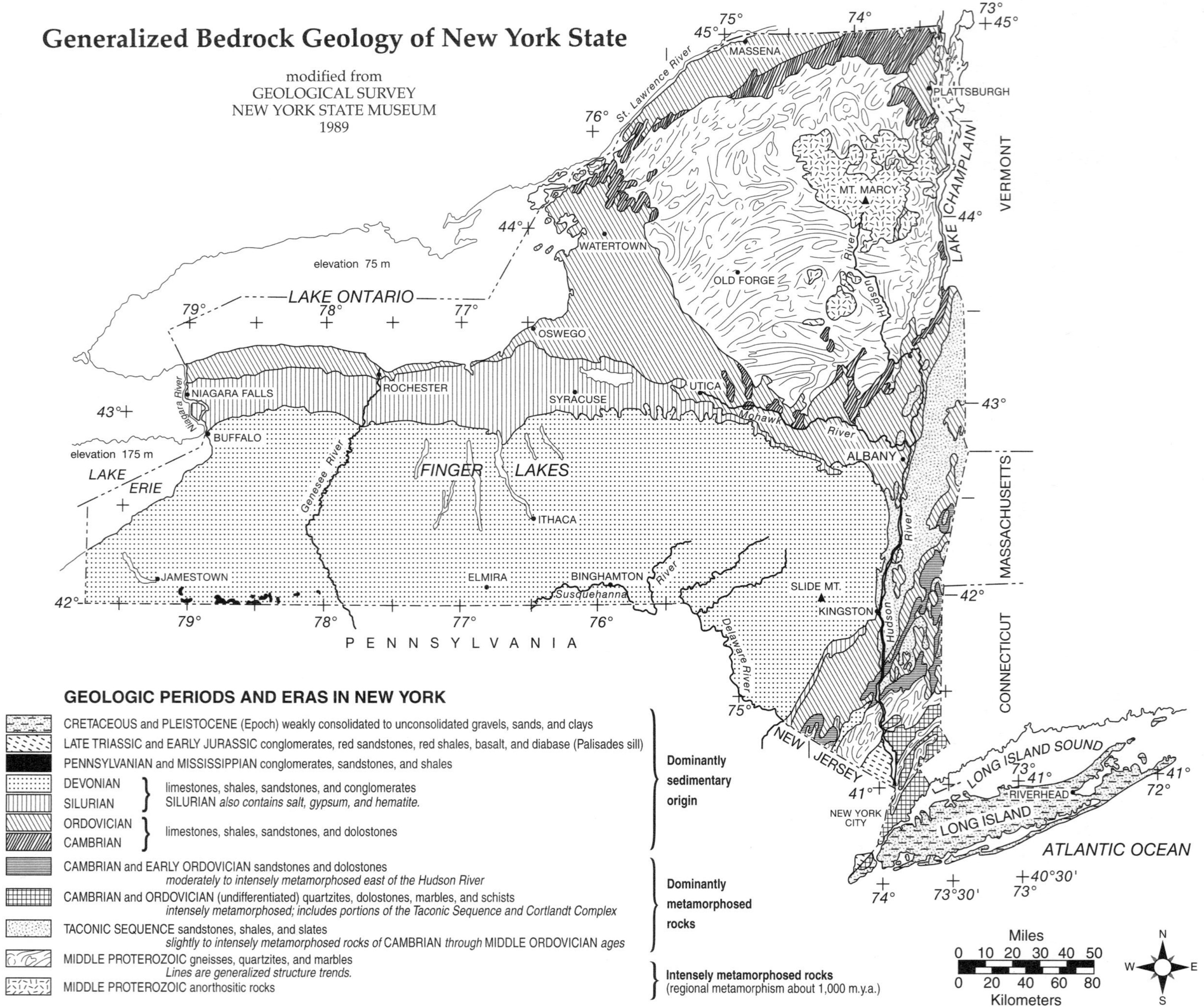

Generalized Landscape Regions of New York State

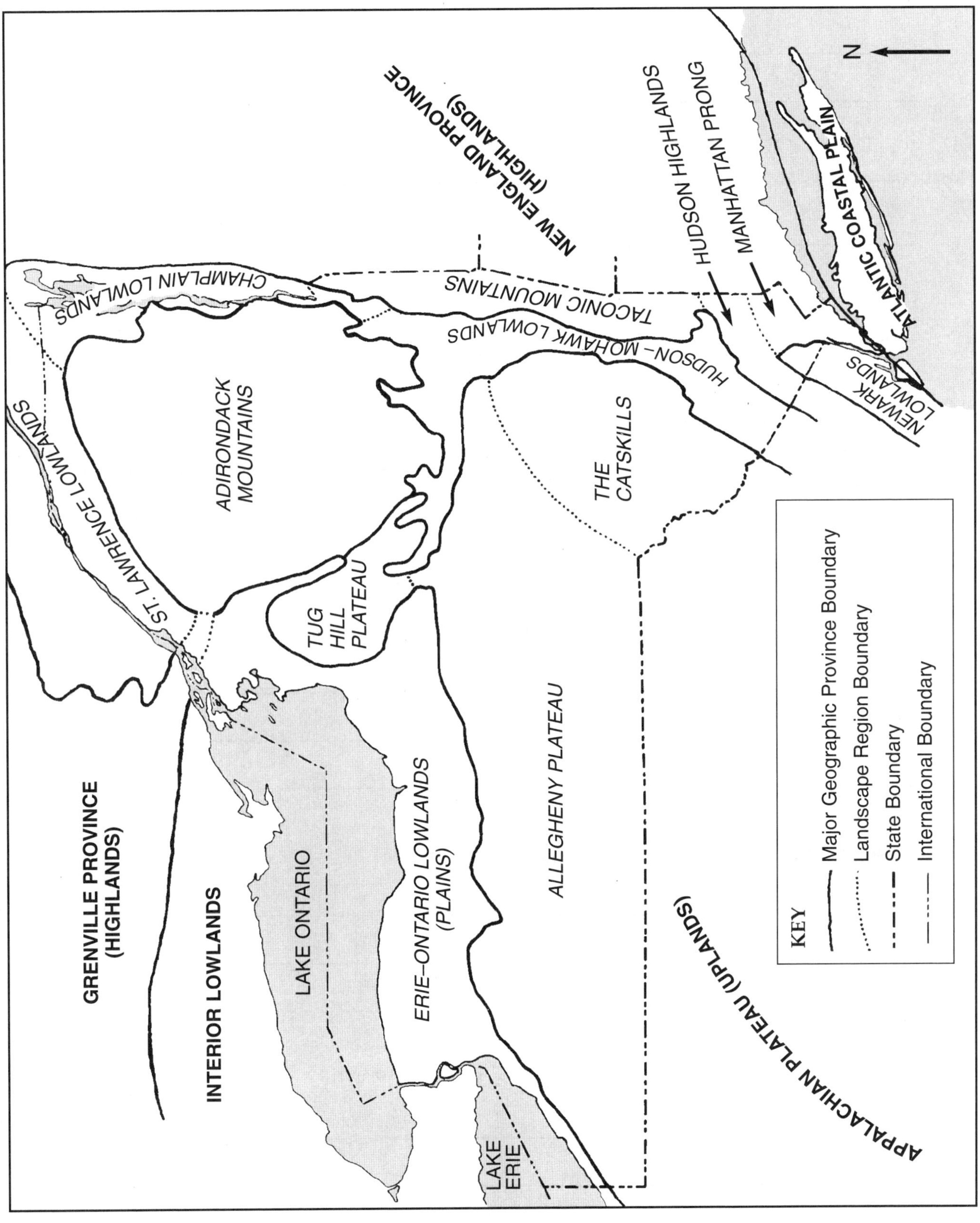

Observations:

1. Describe the type of bedrock in a mountainous region.

2. What landscape region did you describe as mountainous? Why?

3. Describe the type of bedrock in a plateau region.

4. What landscape region did you describe as plateau? Why?

5. Describe the type of bedrock in a plains region.

6. What landscape region did you describe as a plain? Why?

7. What geologic features would one find in a mountain in New York State?

8. Explain why the term "Catskill Mountains" is misleading.

9. Is New York a mountainous state? Explain your answer.

10. Describe the landscape region where you live. Explain and defend your statement citing your local bedrock.

NAME ______________________

20 Local Water

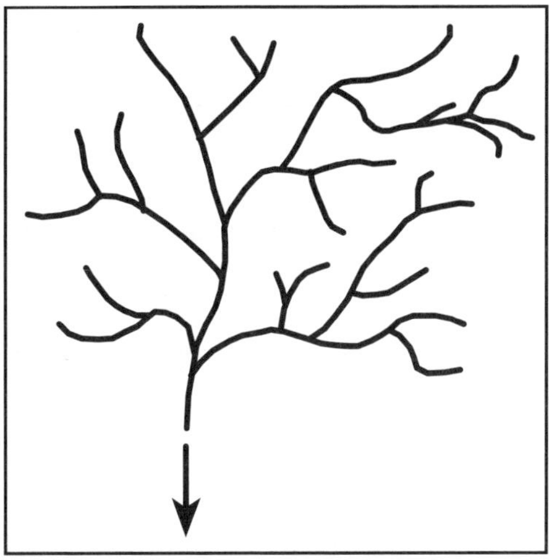

Water is the most precious substance to life on Earth. We must ensure the quality of the water on our planet and our local water supply. Most people on the planet get their drinking water from ground sources because it is naturally purified. Human population brings the need for water but it also brings pollution. Water sources need to be protected from infiltration of pollutants into the water supply. Some places do not have their water underground because of the local geology. Areas that are intensely metamorphosed cannot have ground water infiltration because of the densely packed minerals in the rock. These places must have above surface collection and storage areas of water. They are called reservoirs.

Pre-Lab: Define the following words.

Water cycle: ____

____

Porosity: ____

____

Capillarity: ____

____

Permeability: ____

____

Infiltration: ____

____

Objective:

The objective of this lab is to find out about the quality of drinking water in the local watershed.

Essential Question: ____

____

Materials:
- Computer with internet access
- Environmental Protection Agency site: http://cfpub.epa.gov/surf/locate/index.cfm
- Digital Library for Earth Science Education: http://www.dlese.org

Procedure:

1. Using the above sites, research information about the local watershed by entering your school's zip code.
2. Determine a means to monitor water quality.

Observations:

1. What is the local coverage of the watershed for this area?

   ____

   ____

2. Why is water considered so important?

   ____

   ____

   ____

3. What does geology have to do with where local water is stored?

   ____

   ____

   ____

4. According to the Earth Science Reference Tables, describe the type of bedrock your county has.

5. Is your local watershed an above ground reservoir or an underground aquifer?

Conclusions:

6. What is required to monitor water quality of the local drinking water?

7. What are some substances that can contaminate the drinking water?

8. What can the community do to improve the quality of the drinking water?

9. How can you and your family contribute to protecting your watershed?

____

____

____

____

____

10. What have you learned about the watershed protection today that you didn't already know?

____

____

____

____

____

____

____

____

____

____

____

____

____

NAME ____________________

21 Stream Beds

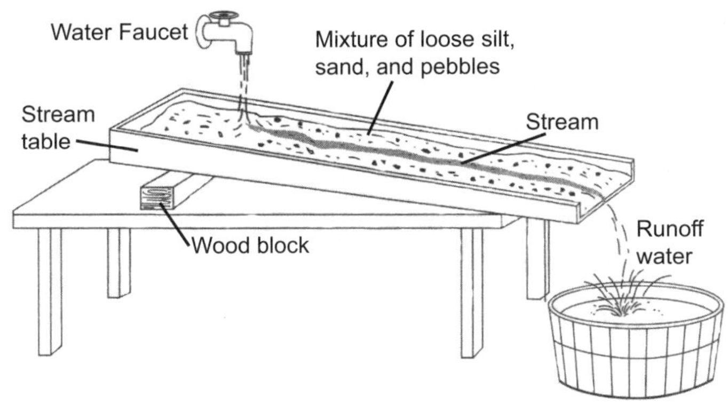

Running water is a major agent of erosion. The stream bed is affected by its shape and the angle in which it flows. The stream flow then carves out the stream channel. The water in the stream travels at a speed dependant on the angle of the slope. The channel in which the stream travels is affected by the slope. Young streams tend to be straight and have steep slopes. As streams mature in their landscapes, the stream will begin to meander. Fully mature streams have flood plains in areas the stream bed gets pinched and forms oxbow lakes.

Objective:

The objective of this lab is to observe streambeds and depositional patterns of streams.

Essential Question:

____

____

Hypothesis:

If____

Then____

Because____

Variables:

Independent:____

Dependent:____

Materials:

- Stream table with adjustable height
- Mixture of gravel and sand
- Water

Procedure 1 - Slight Angle - Straight Channel:

1) Set up the stream table so that it is at a ***slight*** angle.
2) Add the gravel and sand mixture in the middle of the stream table. Smooth out the sediments, producing a slightly sloping landscape. Do *not* put any sediments in the lower end of the stream table.
3) Add water to create a "lake" at the upper end of the stream table. Add water to the lower end to produce a "quite body of water".
4) Make a shallow ***straight*** channel through the sediment mixture.
5) Slowly add water into the lake area so that it flows through the channel to the "quite body of water".
6) Continue adding water for 5 minutes while recording observations.

Observations:

____

____

____

____

Draw the final results of your stream table in the box below.

Procedure 2 - Steeper Angle- Straight Channel:

1) Set up the stream table so that it is at a ***steeper*** angle.
2) Repeat steps 2 through 6 from Procedure 1 – while using the same sediments contained in your stream table.

Observations:

____

____

____

____

Draw the final results of your stream table in the box below.

Procedure 3 - Slight Angle - Meandering Channel:

1) Set up the stream table so that it is at a ***slight*** angle.
2) Repeat steps 2 through 6 from Procedure 1 **except -** Make a shallow ***meandering*** channel through the sediments.

Observations:

____

____

____

____

____

Draw the final results of your stream table in the box below.

Procedure 4 - Evaporation:

Your teacher will select one of the stream tables to observe the results after evaporation.

Observations:

____

____

____

____

____

____

Draw the final results of your stream table in the box below.

Observations:

1. Which stream had the greatest water velocity?

2. Which stream had the deepest channel?

3. What happened to the sides of the channel as far as erosion and deposition?

4. What are some ways in which the stream channel could be adjusted?

Directions: Use the Relationship of Transported Particle Size to Water Velocity chart.

Relationship of Transported Particle Size to Water Velocity

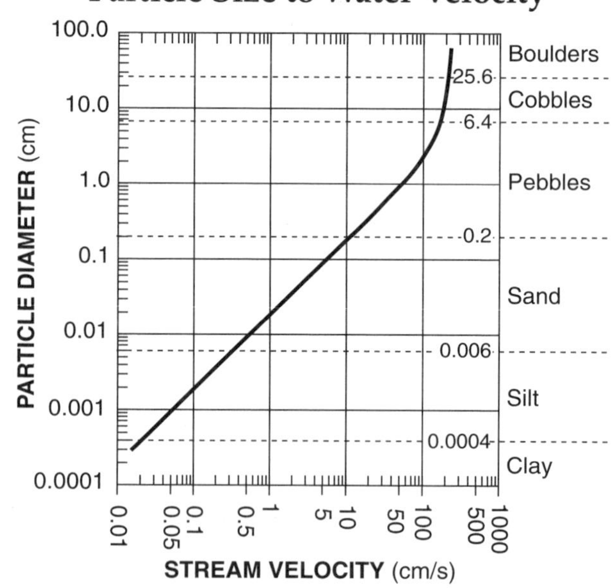

This generalized graph shows the water velocity needed to maintain, but not start, movement. Variations occur due to differences in particle density and shape.

5. What is the minimum stream velocity of water needed to transport the coarse sand?

6. If you had your stream traveling at a velocity of 100 cm per second, what is the largest particle that could be transported?

____

____

____

7. As water slows to still water, what happens to the sediment?

____

____

____

8. In the following stream pattern, please determine the stream bed from *A* to *A′*.

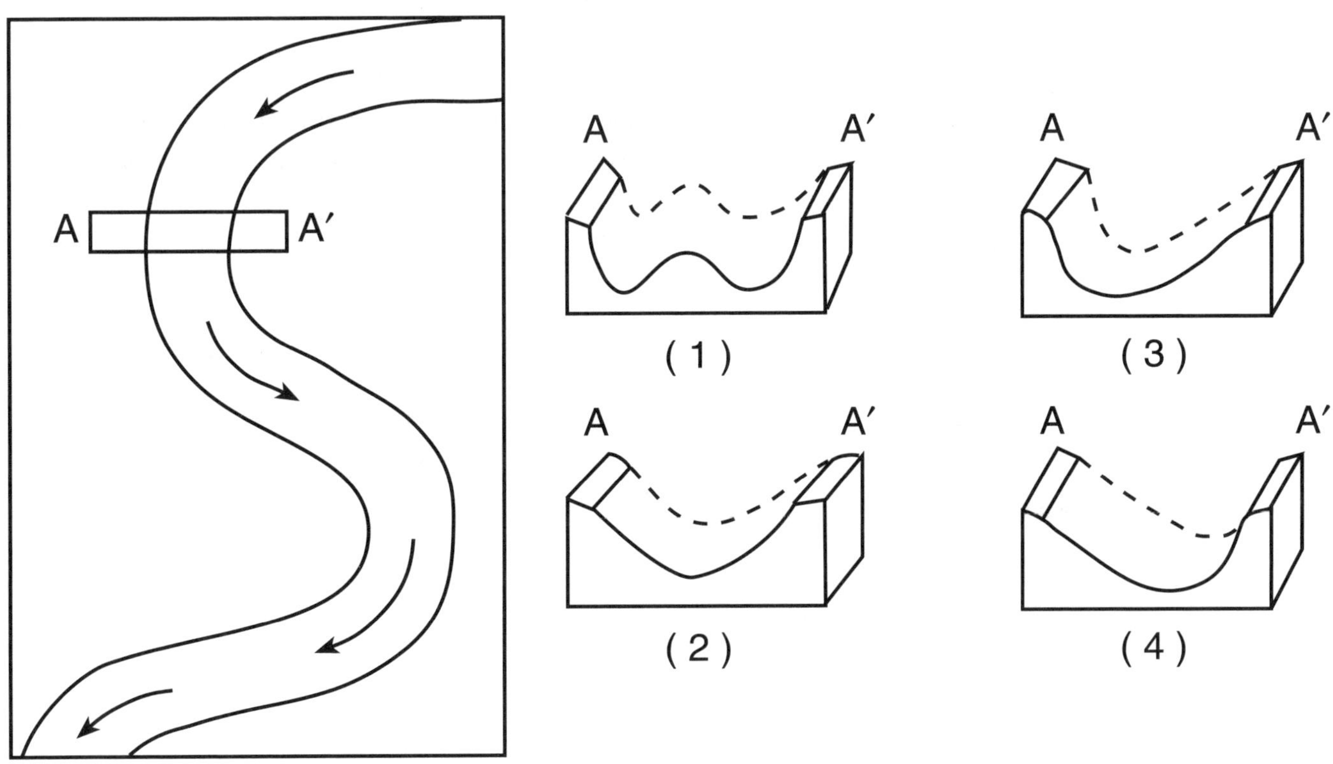

8 ________

9. The illustration below depicts the method in which materials deposit in horizontal sorting.

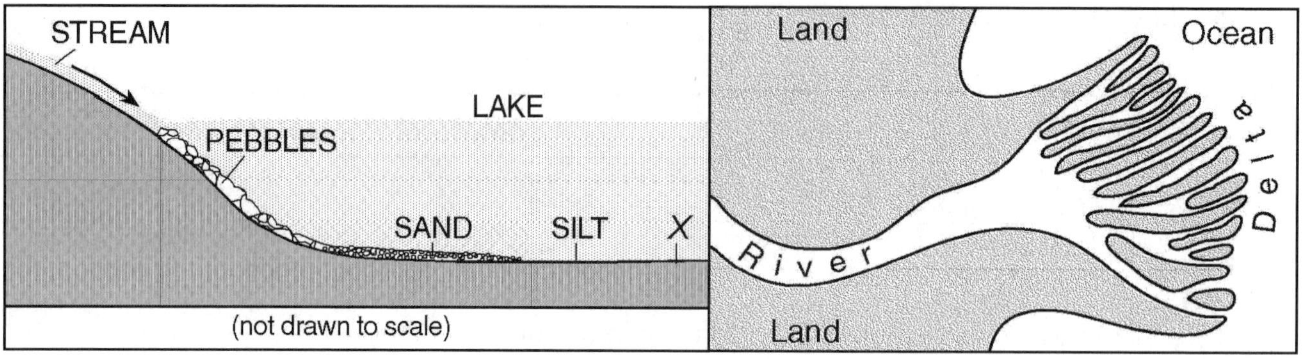

Explain the chart using the Relationship of Transported Particle Size to Water Velocity chart, and use it to explain the formation of the delta.

10. Shown below are three stages of stream development. Think of the three different types of stream channels that you created, compare your streams to the three types below.

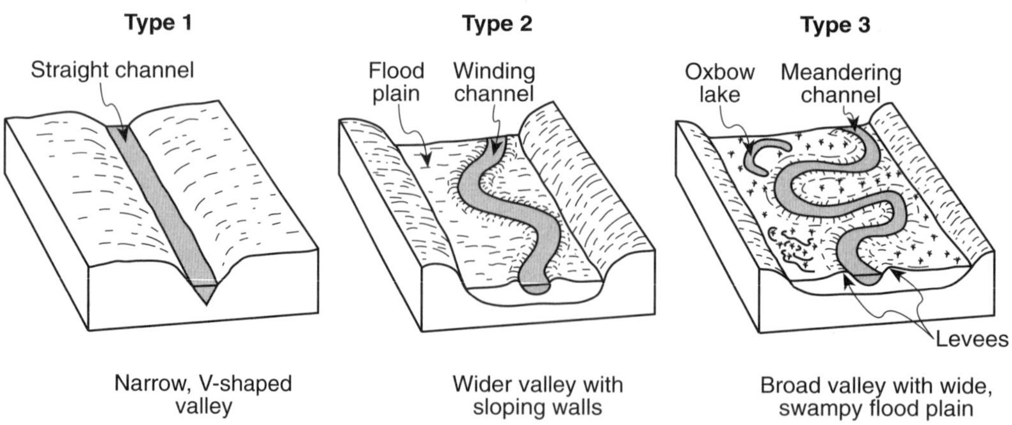

22 Geologic Profiles

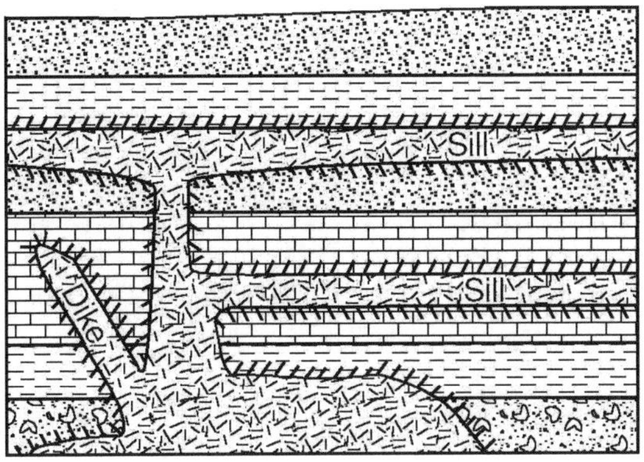

Stratigraphy is the study of the history of the Earth's crust, particularly its stratified rocks. Stratigraphy determines age relationships of rocks as well as their distribution in space and time. Rocks may be studied in an outcrop but commonly are studied from drilled cores. Most of the Earth's surface is covered with sedimentary rocks. These rocks record a geologic history and is what makes stratigraphy important.

Another important feature that geologists look for is the index fossil. An index fossil is the remnant of a once living organism that, as a species, lived for a short period of time and lived over a very wide area.

Sediments, such as those at the bottom of an ocean or lake, are usually deposited in layers. When layers are tilted, it is assumed that the layers were deposited level, and that the area tilted after it turned into sedimentary rock. The rocks used in sequence stratigraphy are bounded by, or surrounded by, surfaces of erosion called unconformities.

"Laws of Stratigraphy"

Law of Cross Cutting Relationships: A rock is always older than the process that affects it; rock strata are older that the faults that cut through them.

Law of Original Horizontality: Sedimentary rocks are almost always deposited in horizontal layers.

Law of Superposition of Strata: Oldest rocks can be found at the bottom of an undisturbed bed.

Law of Uniformitarianism: The process that are currently shaping the Earth have been the same processes since the formation of the Earth; "The present is the key to the past."

Objective:

The objective of this lab is to determine what events might have led to stratigraphic layers.

Essential Question:

____

____

Materials:

Geologic History of New York State Chart in the Earth Science Reference Tables

Procedure:

Using the following pictures, determine the order of the geologic events. Start with the first event, usually deposition, and list the events as they occurred. Defend your findings. Look at the keys to determine the rock types. Use the following terms: deposition, erosion, folding, faulting, intrusion, extrusion.

Look carefully at the clues. Rocks have a story to tell, if you listen.

Consider the example profile blow; let's look at how we can solve the following geologic profile. We see the rock strata are flat so we can apply the Law of Superposition. Layer 1, was deposited first, followed by the deposition of Layer 2. Notice in Outcrop *A*, there are four layers and Outcrop *B* has three layers. Layer 3 is missing, it was eroded away. Then Layer 4 is the youngest. The last event was erosion by a river, we know this because it has a V shaped valley.

Example

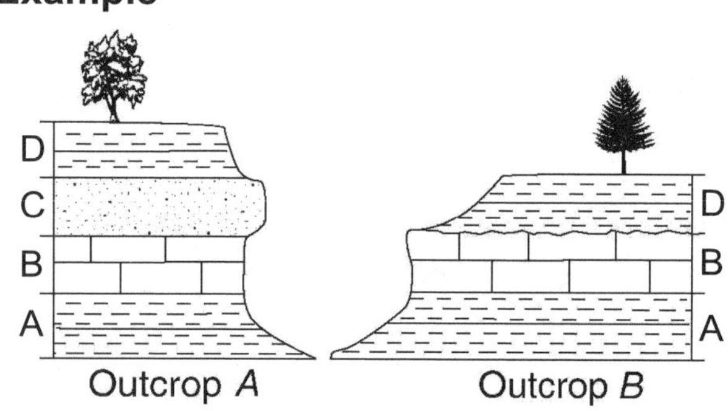

A. Deposition

B. Erosion

C. Deposition

D. Erosion

Directions:

Identify the processes labeled in Profile 1.

Profile 1

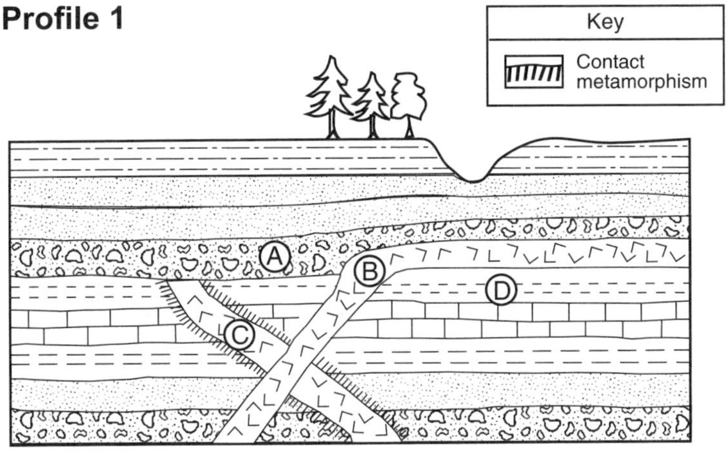

A. ____________________

B. ____________________

C. ____________________

D. ____________________

Directions:

Based on Profile 2 and 3 infer the order in which the Earth processes occurred to form the profile.

Profile 2

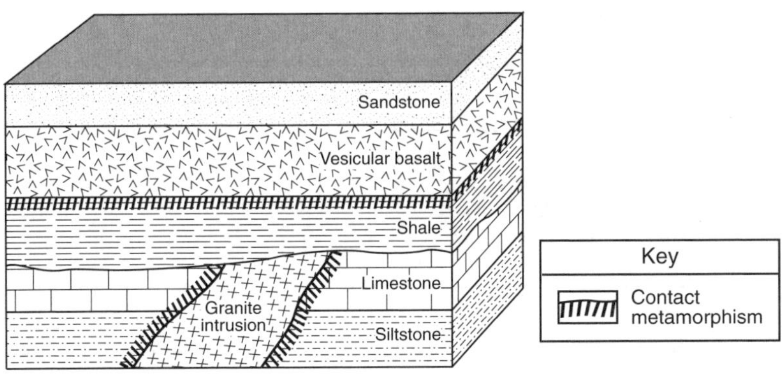

A. ____________________

B. ____________________

C. ____________________

D. ____________________

E. ____________________

F. ____________________

Profile 3

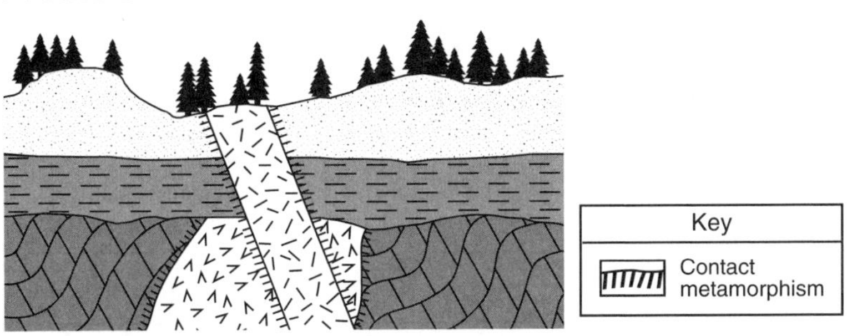

A. ____________________

B. ____________________

C. ____________________

D. ____________________

D. ____________________

E. ____________________

F. ____________________

Observations:

1. In an undisturbed layer, how do we know which layer was deposited first?

2. Why are sedimentary rocks primarily deposited flat?

3. We base these inferences on the earth processes we see today. Upon what assumption do we state that these processes have not changed?

4. Why are sedimentary rocks always almost deposited in flat strata?

5. Why would some parts of layers be missing from the profile?

6. What do you call that missing parts of the rock record?

7. Are these events occurring anywhere in the world today? Explain.

Directions:

Answer the following questions based on the diagram and the chart of the Geologic History of New York in the Earth Science Reference Tables.

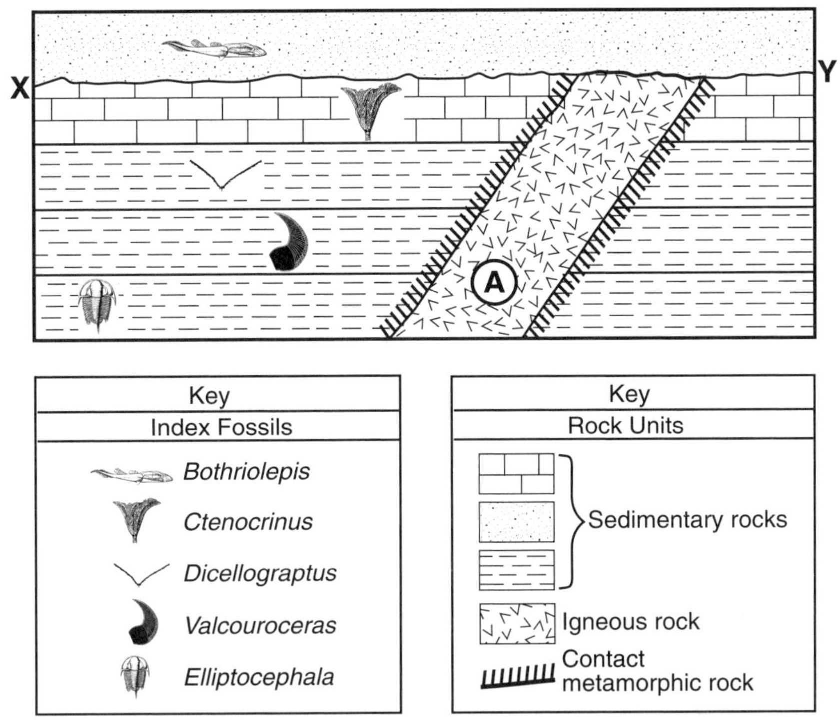

8. Based on the fossil evidence, determine the geologic period during which the unconformity formed.

____

____

9. Each index fossil existed for a relatively short geologic time interval. State one other feature that each fossil must have in order to be considered an index fossil.

____

____

10. Based on the rock type, in what type of environment did the organisms exist?

____

____

____

____

____

____

23 Geologic Calender

NAME ____________________

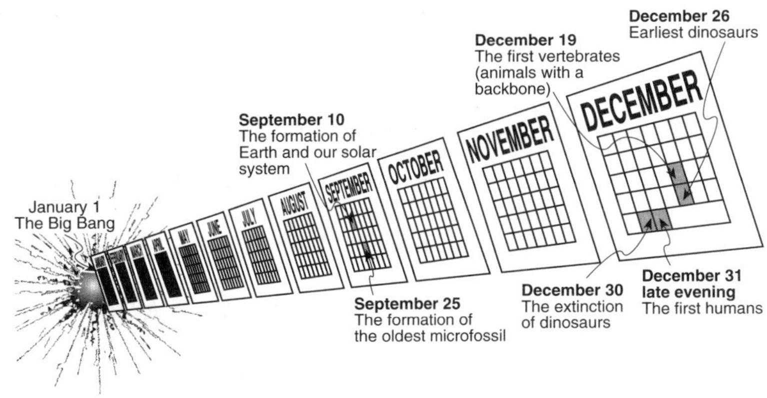

A Geologic Calender is a time scale that covers the Earth's entire geologic history from its origin to the present. Before the growth of a geologic time scale in the 19th century, natural historians recognized that the Earth had a lengthy history. The scale used today was developed over the last 200 years and continues to evolve. A geologic time scale helps scientists think about the history of the Earth in manageable sections of time. The geologic time scale is based on the development of life on Earth.

Objective:

The objective of this lab is to gain an understanding for the vastness of the Earth's past by constructing a geologic calender.

Materials:

- Register tape
- Meter stick
- Crayons or markers
- Masking tape
- Pencil

Procedure:

Below you will find a list of events from the Earth's past. It is your job to find where in the calender the events occurred. You will have to construct 4.6 billion (4,600 million) years of the Earth's history on 5 meters of paper tape.

PROCEDURE:

1. Measure 5 meters of paper tape.
2. Using your pencil, place a vertical line 10 cm from the leading edge. This will be your starting line. This will also be the point where all measurements are made from as you work back in time.

Scale: 1 cm = 10 million years

3. Using the color key below- divide the time line into the eras.

Precambrian Era	Blue diagonal stripes
Paleozoic Era	Green diagonal stripes
Mesozoic Era.	Orange diagonal stripes
Cenozoic Era.	Pink diagonal stripes

4. In the same colors, write the name of each era.
5. Place all lab partners' names in the 10 cm space by the start of the time line.

Pre-Lab: Calculate centimeters from present for each of the following events

EVENT:	**YEARS AGO:**	**Scale: 1 cm = 10 million years**
Formation of the Earth	4.6 billion (4600 million)	**460 cm**
Oldest rocks	4.3 billion (4300 million)	
First plant (algae)	3.2 billion (3200 million)	
Abundant stromatolites	1.3 billion (1300 million)	
First animal (jellyfish)	580 million	
Start of Paleozoic	**542 million**	
First vertebrate animal	500 million	
First land plants	440 million	
First insects	420 million	
First reptiles	350 million	
Start of Mesozoic	**251 million**	
First dinosaurs	225 million	
First mammals	210 million	
Break-up Pangaea	200 million	
First birds	150 million	
Dinosaurs extinction	66 million	
Start of Cenozoic	**65.5 million**	
First man-like animals	4 million	
Ice Ages start	1 million	
First Neanderthal man	0.11 million	
Today	0.00 million	

Observations:

1. Which is the longest era?

2. What is the New York State fossil?

3. Humans and dinosaurs did not live at the same time. What amount of time separated them?

4. Explain your answer to question number 3 by citing what you've learned in this lab.

5. How did the scientists divide the time periods?

6. Where do you personally fit into the time scale?

7. When did the rocks in the Adirondacks begin to metamorphose?

8. How did the Palisades Sill form?

9. During what period did the New York State fossil live?

10. What is the earliest appearing animal whose species is still alive today?

NAME ____________________

24 M & M® Half Life Lab

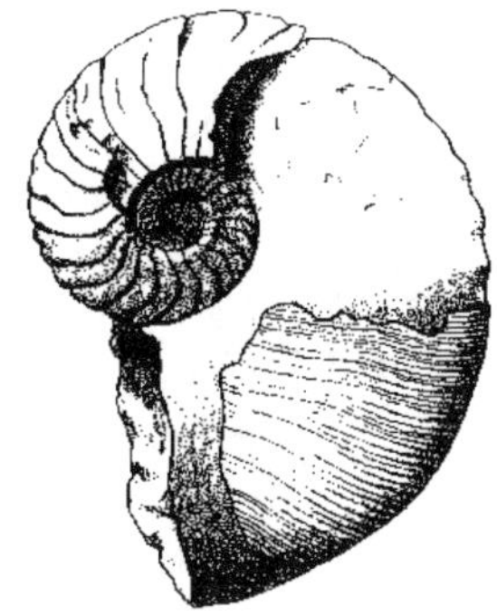

Centroceras

Radioactivity is the natural breakdown of atomic nuclei. In 1898 the French chemists Marie Curie and Pierre Curie figured out that radioactivity is an occurrence associated with atoms, regardless of their physical or chemical state.

When the Sun's rays hit our planet, some particles react with the nitrogen in the atmosphere. The atom of nitrogen becomes changed, and once a neutron is gained, a proton is lost. One can no longer call this atom nitrogen. It is called $Carbon^{14}$

The decay of some substances, such as $Potassium^{40}$ and $Carbon^{14}$, appear to continue indefinitely with a regular decay rate. Each individual radioactive substance has a unique half-life. A half-life is the amount of time it takes the radioactive material to decay to a stable product.

Objective:

The objectives of this lab is to demonstrate radioactive decay.

Essential Question:

____

____

Hypothesis:

If ____________________

Then ____________________

Because ____________________

Variables:

Independent: ____________________

Dependent: ____________________

Materials:

- 100 M & M® candies
- Box
- Use the accompanying chart to help answer questions in the Observation section.

Radioactive Decay Data

RADIOACTIVE ISOTOPE	DISINTEGRATION	HALF-LIFE (years)
Carbon-14	$^{14}C \rightarrow {}^{14}N$	5.7×10^{3}
Potassium-40	$^{40}K \rightarrow {}^{40}Ar$, $^{40}K \rightarrow {}^{40}Ca$	1.3×10^{9}
Uranium-238	$^{238}U \rightarrow {}^{206}Pb$	4.5×10^{9}
Rubidium-87	$^{87}Rb \rightarrow {}^{87}Sr$	4.9×10^{10}

Procedure:

1. Place the M & M candies with the “M” facing up in the bottom of the shoebox. This is the starting number.
2. Close the cover and <u>gently</u> shake.
3. Open the box and remove all the changed candies - those that now have the “M”-side down.
4. Count and record the number of unchanged candies remaining in the box (those with the M’s up)
5. *Record this on data chart. Also record the number of changed candies.*
6. Repeat steps 2 through 5 until all the candies have turned.
7. Graph the information from the chart. In this model of half-life decay - The X-axis will be the number of shakes (independent variable). On the Y-axis, plot the number changed in red, plot the number unchanged in blue.

Shake	**0**	**1**	**2**	**3**	**4**	**5**	**6**	**7**	**8**	**9**	**End**
number of unchanged	100										0
number of changed	0										100

Title: ______________________________

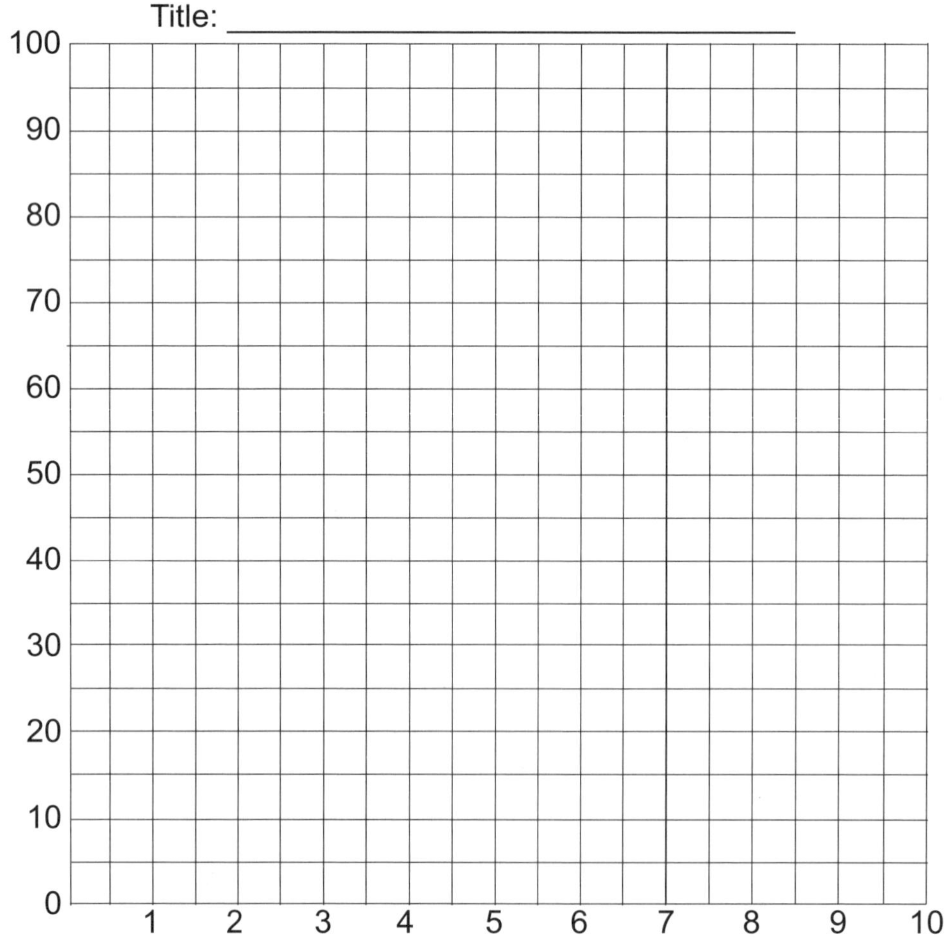

Observations:

1. How many 'atoms' changed by the end of this lab?

____

____

2. Compared to the original number of atoms you started with approximately what percentage was left after each trial?

____

____

3. If each time the box gets shaken represents 500 years, what would the half-life of the atoms (M & M) be?

____

____

4. What is the half-life of $Carbon^{14}$?

____

____

5. Where does $Carbon^{14}$ in nature come from? Explain.

____

____

____

____

6. What are some problems of this experiment in demonstrating half-life?

____

____

7. Can $Carbon^{14}$ dating be used for dating organic substances from the Precambrian era? Explain.

____

____

8. Can Carbon^{14} be used to date quartz? Explain.

____

____

9. If you have a sample of Calcium^{40} as a decay product, what would the parent material have been?

____

____

10. If you currently have 1 Kg of Uranium^{238}, in how many in years will you have ½ Kg of Lead^{206}?

____

____

____

____

NAME ____________________

25 Coriolis Effect

The rotation of Earth from a western to an eastern direction causes free-moving objects, such as air and water, to move at different rates. This is called the Coriolis Effect. For example, in the northern hemisphere, air moving from the North Pole toward the equator is deflected to the right. The opposite is true for the Southern Hemisphere. The Coriolis Effect influences the movement of global wind patterns and ocean currents.

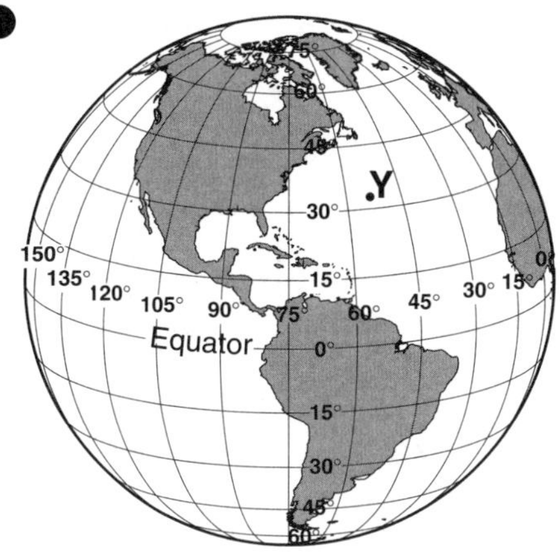

Objective:

The objective of this lab is to illustrate planetary wind motion and to find patterns of ocean currents.

Materials: • Ruler • Planetary map • Reference Tables

Pre-lab:

Answer the following questions using the Surface Ocean Currents chart and the Planetary Wind and Moisture Belts in the Troposphere diagram. These charts can be found in the back of your lab book.

1. You are in a classroom between 40°30’ N and 45°N. What global wind pattern are you in?

____

2. If there was a “change in the weather”, where would it most likely track from?

____

3. Suppose you were in a classroom in Miami, in southern Florida. It is south of 30°N. What would your global wind pattern be?

____

4. From your classroom in Florida, which direction would the weather generally track from?

____

Procedure:

Use the grid below of the planet Earth. Note that longitude is referenced on both top and bottom of map. Follow the directions and complete questions as listed below.

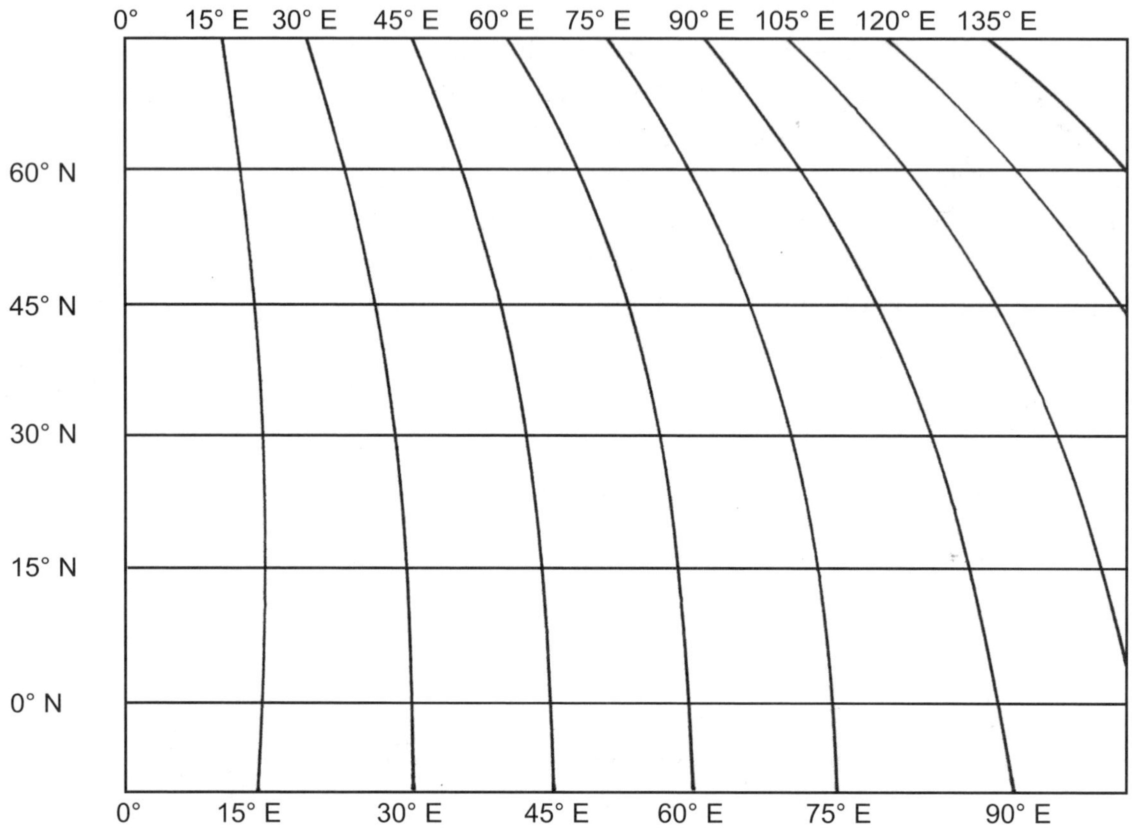

1. You live in the imaginary community of Ersatz at 15° N and 90° E. With your red pencil, draw a dot and circle the point like this ⊙ and label it.
2. Your community has a trade relationship due north of you in the town of Bogus at 45° N and 90° E. With your red pencil, draw a dot and circle the point like this. ⊙
3. Trace the path, with your red pencil, that one would take to go to this route in a dashed line (- - - - - - -).
4. Over time, your technology has progressed. You can send items via a projectile. These projectiles travel so fast, that people in these communities did not take into consideration the Earth's rotation. Draw a dashed line, with your red pencil, straight north of your home community. The line should be parallel to the edge of the paper.
5. Now you are visiting your distant cousin in Faux. Faux is not tropical like Ersatz. Faux is 60° N and 90° E. With your green pencil, draw a dot and circle the point like this ⊙ and label it.
6. Your cousin from Faux has decided to be your trading partner. Your cousin is excited to get the items to you quickly so he sends the items due south. Draw a dashed line, with your green pencil, straight south from Faux. The line should be parallel to the edge of the paper.
7. The people of Ruse are delighted with the delivery.

Observations:

1. Looking at the error on the map on the previous page, what is the difference in degrees, between what should have been the target, Bogus, and what was the target by the projectile?

____

2. From the point of view of the residents of Ersatz, did the projectile "curve" to the right or the left? ____________________

3. The projectile landed in the city of Ploy. What are the terrestrial coordinates for Ploy?

____

4. From the point of view of the residents of Faux, did the projectile "curve" to the right or the left? (The point of view looking the opposite way of Ersatz, turn your paper around) __________

5. The projectile landed in the village of Ruse. What are the terrestrial coordinates for Ruse?

____

6. Look at the the Planetary Wind Belt map in the Earth Science Reference Tables on page 216. The planetary wind belts of the SW winds are traveling in what direction?

____

7. Do the winds curve to the left or to the right? ____________________

8. Look at the the Surface Ocean Currents map in the Earth Science Reference Tables. Look at the Gulf Stream off the eastern coast of North America. Describe its direction of travel, its apparent curve, etc.

____

____

9. Look at the the Surface Ocean Currents diagram in the Earth Science Reference Tables on page 216. Look at the California Current off the western coast of North America. Describe its direction of travel, its apparent curve, etc.

10. Why do the hydrosphere and atmosphere seem to not be moving at the same speed as the lithosphere?

NAME ______________________

26 Atmospheric Pressureand Cyclones

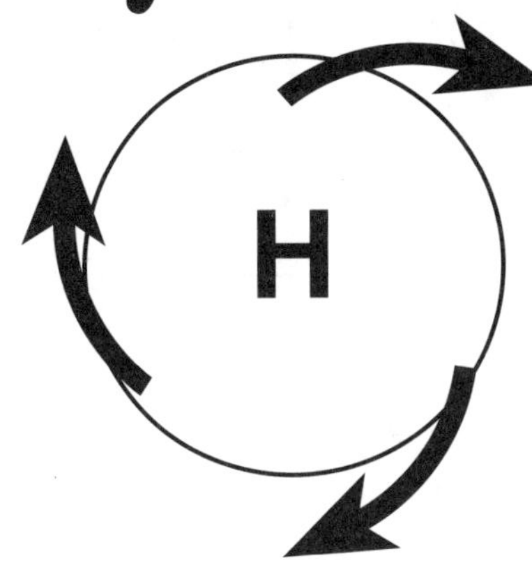

The atmosphere is made up of certain gases. The gases have mass. Because gas has mass it will exert pressure. Atmospheric gases often confuse people because they generally can't be seen or felt. We will be working with atmospheric water as a gas, not as a liquid. We now need to think of water as vapor, a gas.

This investigation explores atmospheric density, pressure and motion. We will determine the density of the differing air qualities. We will be looking at the qualities of the air masses. Cyclones are zones of convergence. This means that winds are coming together from different directions. Anti-cyclones are divergent areas, which behave in an opposite manner. In anti-cyclones winds go in separate directions.

Complete the chart on the next page and determine the atomic weight of the atmospheric gases. Use your reference table to help you. We will use a simulator to demonstrate the two qualities of air as they move pass each other.

Objective:

The objective of this investigation is to compare the atmospheric mass of dry and moist air, the relationship of the mass to pressure, and atmospheric movement.

Essential Question:

____

____

Hypothesis:

If ____

Then ____

Because ____

Variables:

Independent: ____

Dependent: ____

Materials:
- Data chart
- Calculator
- Graph grid
- Paint brush
- Tornado tube
- Earth Science Reference Tables (back of Lab Handbook)

Procedure:

Listed below, but not in order, are the gases of the atmosphere. Using the data below, calculate the molecular mass of the molecules in Chart 1.

Atom:	Carbon	Hydrogen	Oxygen	Nitrogen
Mass:	12	1	16	14

CHART 1

Molecule	**Molecular Formula**	**Molecular Mass**
Carbon Dioxide – CO_2	1 carbon, 2 oxygen	
Oxygen – O_2	2 oxygen	
Nitrogen – N_2	2 nitrogen	
Water Vapor – H_2O	2 hydrogen 1 oxygen	

Complete Chart 2. This chart will yield the atomic weight of 100 molecules of air with absolutely no water vapor in it. Use Chart 4 on the next page to assist you. Use CO_2 as 1%.

CHART 2

1 **Name of Gas**	**2** **% of Gas in Atmosphere**	**3** **(Multiply)**	**4** **Molecular Mass ***	**5** **Total**
Carbon Dioxide	1	X		1.
Nitrogen		X		2.
Oxygen		X		3.
			TOTAL	4.

*** Use the molecular mass from Chart 1 in the other two charts.**

Water vapor in the air is at a maximum of 5%. We still want to measure the weight of 100 molecules (100%) so we need to remove the correct amount of molecules. Fill in the chart below, using the data from the previous chart. Subtract column 3 from column 2. Round to the *nearest whole number* and place your answer in column 4. Multiply column 4 by column 5.

CHART 3

1 Name of Gas	2 % of Gas in the Atmosphere	3 Multiply by .05	4 Subtract column 3 from 2	5 Molecular Mass*	6 TOTAL
Carbon Dioxide			X	=	
Nitrogen			X	=	
Oxygen			X	=	
Water Vapor	5%		X	=	
				TOTAL	

CHART 4

Average Chemical Composition of Earth's Crust, Hydrosphere, and Troposphere

ELEMENT (symbol)	CRUST		HYDROSPHERE	TROPOSPHERE
	Percent by mass	Percent by volume	Percent by volume	Percent by volume
Oxygen (O)	46.10	94.04	33.0	21.0
Silicon (Si)	28.20	0.88		
Aluminum (Al)	8.23	0.48		
Iron (Fe)	5.63	0.49		
Calcium (Ca)	4.15	1.18		
Sodium (Na)	2.36	1.11		
Magnesium (Mg)	2.33	0.33		
Potassium (K)	2.09	1.42		
Nitrogen (N)				78.0
Hydrogen (H)			66.0	
Other	0.91	0.07	1.0	1.0

*** Use the molecular mass from Chart 1.**

Observations:

1. Which has a greater mass, dry air or moist air?

____

____

2. Determine the densities of dry air and moist air, assuming that the measured air occupies 1 cm^3. (*Show all your work.*)

3. Knowing what we already know about density differences in Earth systems, what do you predict the less dense air will do in a relation to more dense air?

____

____

____

____

4. If something is pushing down, it is exerting pressure. Which air quality, dry or moist will have higher pressure?

____

____

5. When one air quality has a greater density its pressure pushes down. Take the paintbrush in your hand with the bristles just touching the table. Gently exert force on it. Notice what the bristles do. Relate the behavior to what you know about wind patterns. (*Look to investigation 25 for reference.*)

____

____

____

____

Planetary Wind and Moisture Belts in the Troposphere

The drawing on the right shows the locations of the belts near the time of an equinox. The locations shift somewhat with the changing latitude of the Sun's vertical ray. In the Northern Hemisphere, the belts shift northward in the summer and southward in the winter.

(Not drawn to scale)

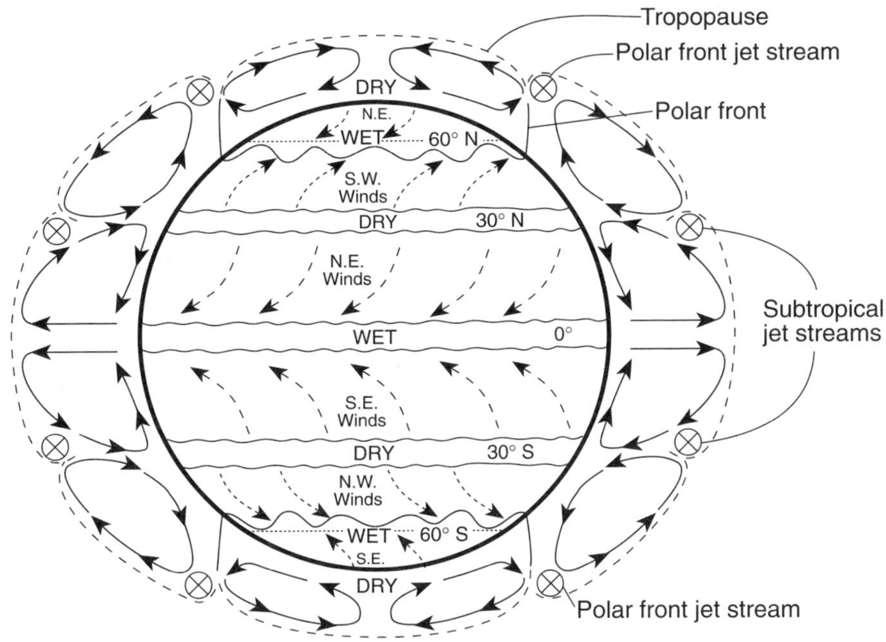

6. In the accompanying box with an **H** in the center, draw how you might think the air mass will behave. Use arrows to indicate the motion of the air and the direction it would flow.

Infer: Look at the accompanying drawing. Knowing how high pressure behaves vertically, and moves in a direction as illustrated in the chart above, use your right hand starting with all of your fingers touching the H, and touching with your fingertips only, mimic the motion of the arrows, in the chart.

H

7. Undo the motion you just did. In the accompanying box with **L** in the center, illustrate the motion using arrows to show the path the air would take and arrows showing direction.

L

Procedure:

Take a tornado tube. Turn the tube so that the water is in the upper bottle. Let the water fall without any movement on your part. Set down and observe.

8. Describe what you see.

____

____

____

____

____

Now repeat the process but this time hold the bottle on the desk with one hand. Move the top of the bottle in quick circular motion.

9. Describe what you see.

____

____

____

____

____

10. In this investigation, water in the atmosphere has been discussed in the form of gas. Please describe what the water and air in the bottle represents in relation to what you've been investigating.

____

____

____

____

____

NAME ____________________

27 Dewpoint and Humidity

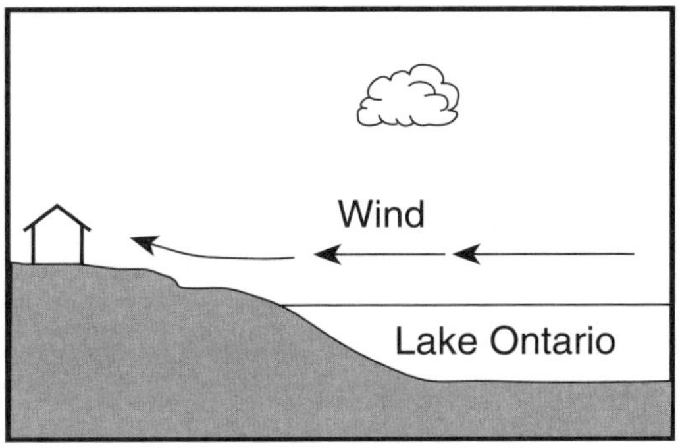

(Not drawn to scale)

Humidity is the moisture content of the atmosphere. The troposphere always contains some moisture in the form of water vapor. The amount of water vapor that will saturate the air increases with a rise in temperature. When the atmosphere is saturated with water, it becomes uncomfortable. To find the relative humidity of the atmosphere, you use a humidity chart. You will also need to know the temperature of a dry-bulb thermometer as well as the difference in temperature between the dry and wet thermometers. Humidity is measured using a psychrometer.

Objective:

The objective of this lab is to determine relative humidity and dewpoint in the atmosphere at a given time.

Essential Question:

____

____

Hypothesis:

If ____

Then ____

Because ____

Variables:

Independent: ____

Dependent: ____

Specific humidity is the actual amount of water vapor in the air expressed in units of mass. Changes in temperature or the amount of water vapor in the air will cause the relative humidity to change.

Graph the information from the chart to the right on the graph below. Place an **X** for each point that you graph.

Amount of Water Vapor That Will Saturate 1 Cubic Meter of Air at Different Temperatures

Air Temperature (°C)	**Water Vapor** (g/m^3)
–20	1
–10	2
0	5
10	9
20	17
30	29
40	50

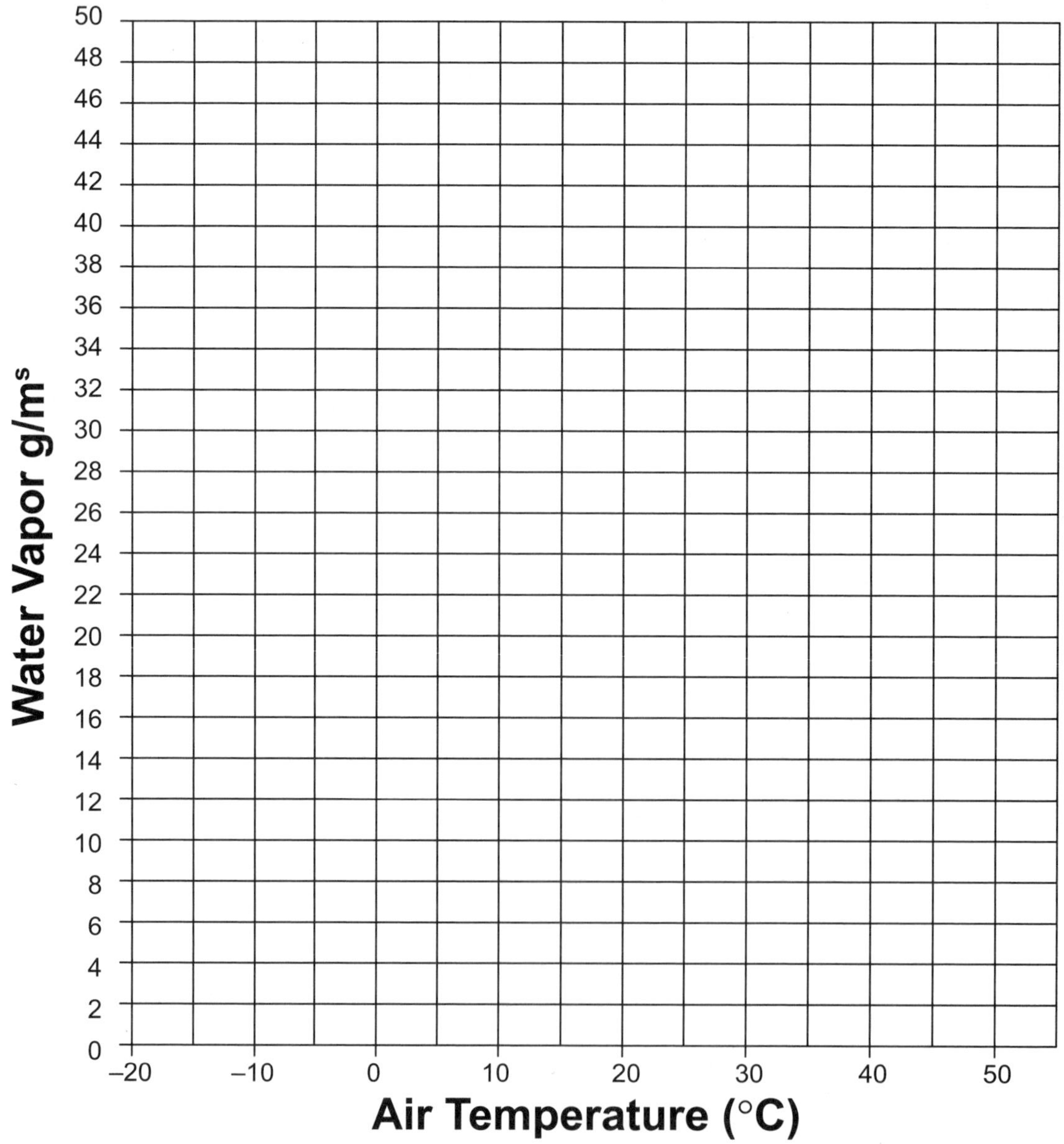

Relative humidity is the amount of water vapor in the air compared to the amount of water vapor the air can hold at capacity.

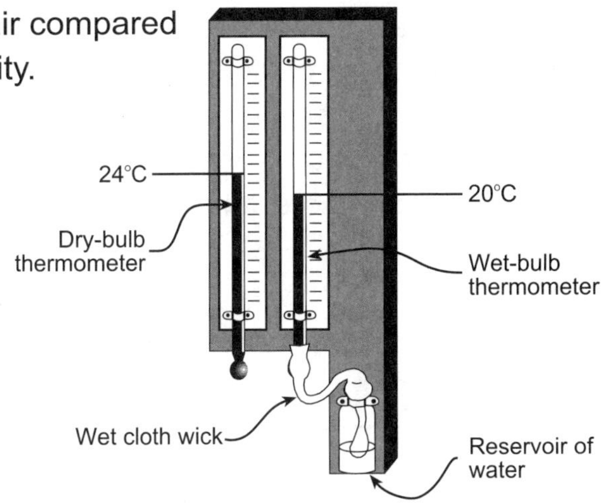

Materials: • Sling psychrometer

Procedure:

1. Wet the wick on the sling psychrometer.
2. Rotate the psychrometer in a rapid motion. First spin for 1 minute. Then spin for 2 minutes. Record your findings.
3. Once you have recorded the indoor readings, report to your teacher, then go outside to do the second set of readings.

Relative Humidity (%)

Dry-Bulb Temperature (°C)	Difference Between Wet-Bulb and Dry-Bulb Temperatures (C°)															
	0	1	2	3	4	5	6	7	8	9	10	11	12	13	14	15
−20	100	28														
−18	100	40														
−16	100	48														
−14	100	55	11													
−12	100	61	23													
−10	100	66	33													
−8	100	71	41	13												
−6	100	73	48	20												
−4	100	77	54	32	11											
−2	100	79	58	37	20	1										
0	100	81	63	45	28	11										
2	100	83	67	51	36	20	6									
4	100	85	70	56	42	27	14									
6	100	86	72	59	46	35	22	10								
8	100	87	74	62	51	39	28	17	6							
10	100	88	76	65	54	43	33	24	13	4						
12	100	88	78	67	57	48	38	28	19	10	2					
14	100	89	79	69	60	50	41	33	25	16	8	1				
16	100	90	80	71	62	54	45	37	29	21	14	7	1			
18	100	91	81	72	64	56	48	40	33	26	19	12	6			
20	100	91	82	74	66	58	51	44	36	30	23	17	11	5		
22	100	92	83	75	68	60	53	46	40	33	27	21	15	10	4	
24	100	92	84	76	69	62	55	49	42	36	30	25	20	14	9	4
26	100	92	85	77	70	64	57	51	45	39	34	28	23	18	13	9
28	100	93	86	78	71	65	59	53	47	42	36	31	26	21	17	12
30	100	93	86	79	72	66	61	55	49	44	39	34	29	25	20	16

Data:

Chart 1	**Dry Bulb (°C)**	**Wet Bulb (°C)**	**Difference (subtract)(°C)**	**Relative Humidity (%)**
Indoors 1 mins.				
Indoors 2 mins.				
Outdoors 1 mins.				
Outdoors 2 mins.				

** Remember to subtract the wet bulb from the dry bulb.*

Dew is water that condenses on cooler objects from the water vapor in warmer air. Air can contain only a certain amount of water vapor. This amount increases as the temperature rises and decreases as the temperature drops. In the evening after a warm day, air that is nearly saturated with vapor cools and drops below the temperature at which it is fully saturated. As the air cools more, the excess water vapor condenses on any surface, such as a blade of grass.

Dewpoint (°C)

Dry-Bulb Temperature (°C)	**Difference Between Wet-Bulb and Dry-Bulb Temperatures** (C°)															
	0	**1**	**2**	**3**	**4**	**5**	**6**	**7**	**8**	**9**	**10**	**11**	**12**	**13**	**14**	**15**
−20	−20	−33														
−18	−18	−28														
−16	−16	−24														
−14	−14	−21	−36													
−12	−12	−18	−28													
−10	−10	−14	−22													
−8	−8	−12	−18	−29												
−6	−6	−10	−14	−22												
−4	−4	−7	−12	−17	−29											
−2	−2	−5	−8	−13	−20											
0	0	−3	−6	−9	−15	−24										
2	2	−1	−3	−6	−11	−17										
4	4	1	−1	−4	−7	−11	−19									
6	6	4	1	−1	−4	−7	−13	−21								
8	8	6	3	1	−2	−5	−9	−14								
10	10	8	6	4	1	−2	−5	−9	−14	−28						
12	12	10	8	6	4	1	−2	−5	−9	−16						
14	14	12	11	9	6	4	1	−2	−5	−10	−17					
16	16	14	13	11	9	7	4	1	−1	−6	−10	−17				
18	18	16	15	13	11	9	7	4	2	−2	−5	−10	−19			
20	20	19	17	15	14	12	10	7	4	2	−2	−5	−10	−19		
22	22	21	19	17	16	14	12	10	8	5	3	−1	−5	−10	−19	
24	24	23	21	20	18	16	14	12	10	8	6	2	−1	−5	−10	−18
26	26	25	23	22	20	18	17	15	13	11	9	6	3	0	−4	−9
28	28	27	25	24	22	21	19	17	16	14	11	9	7	4	1	−3
30	30	29	27	26	24	23	21	19	18	16	14	12	10	8	5	1

1. Fill in the the first three columns using the same data as in Chart 1 on page 131.
2. Use the information from the Dewpoint table above to complete Chart 2 below.

Chart 2	**Dry Bulb (°C)**	**Wet Bulb (°C)**	**Difference (subtract)(°C)**	**Dewpoint (°C)**
Indoors 1 mins.				
Indoors 2 mins.				
Outdoors 1 mins.				
Outdoors 2 mins.				

Observations:

1. Which can hold more water vapor in the air, warm air or cold air?

2. What is the saturation level or water vapor when the air is room temperature, 20°C?

3. Why do summer weather forecasts call for humidity more than winter forecasts do?

4. Based on your findings, which has greater humidity, indoors or outdoors? Why?

5. Using the indoor results for humidity, was there a difference in readings between the 1 minute spin and the 2 minute spin? Explain.

6. Which has a greater dewpoint, indoors or outdoors? Why?

7. Using the indoor results for dewpoint, was there a difference in readings between the 1 minute spin and the 2 minute spin? Explain.

____

____

____

8. Describe the factors that contributed to the humidity being different in the two places you sampled.

____

____

9. Looking at the dewpoint, what indicators can you describe about the current weather?

____

____

10. As the the dewpoint temperature and the air temperature get closer together, what happens to the relative humidity?

____

____

NAME ______________________

28 Clouds

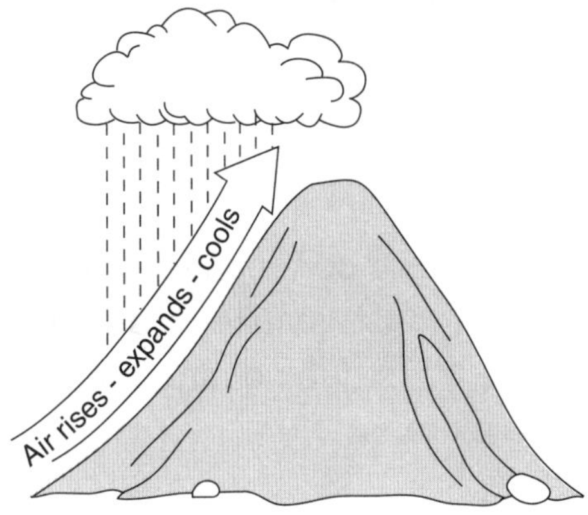

A cloud is a condensed form of atmospheric moisture made up of small water droplets or tiny ice crystals. We are familiar with clouds when we step out of the shower and have a fogged mirror. The mirror is cooler than the air so the water vapor condenses onto that surface.

The cloud formation is caused by air cooling below the dewpoint. Invisible water vapor produces visible cloud droplets or ice particles. Cloud particles are so small and light that upward currents easily carry them in the air. The different cloud formations result partly from the temperature at which condensation takes place. When air rises, it expands because of lessening pressure. Expanding air cools. This is called adiabatic expansion. When air descends, it gets compressed and heats. This is called adiabatic compression. Because clouds will only form where the water vapor can condense and when condensation nuclei are present, the clouds must form above the elevation/altitude that is below the dewpoint. When a cloud falls below that line, the water droplets evaporate. This is why some clouds appear to be flat bottomed.

Objective:

The objective of this lab is to determine how clouds form.

Essential Question:

____

____

Hypothesis:

If ____

Then ____

Because ____

Variables:

Independent: ____

Dependent: ____

Materials:
- 1 liter plastic soda bottle
- Matches
- Water

Safety Precautions:

Flame used. Hair should be tied back, loose clothing secured. No horseplay!

Part 1:

Water vapor condenses onto tiny particles called condensation nuclei. Condensation nuclei can be salt, dust, smoke or any other particulate in the atmosphere, even airplane exhaust. The white trails behind airplanes are actually clouds that formed as water vapor condenses onto the exhaust.

Procedure:
(This event happens quickly, so pay close attention.)

1. Pour 50 ml of water into a one liter clear plastic bottle.
2. Swish water around in the bottle so that bottle is wet inside and there is water at the bottom of bottle.
3. Either you or your partner gently squeeze the bottle. Have the cap ready.
4. The other person will strike a match and then blow out the flame directly over the mouth of the bottle.
5. The bottle should be simultaneously let out so that the smoke is sucked into the bottle. Drop the match in.
6. Cap the bottle quickly.
7. Observe the interior of the bottle.

Figure 1

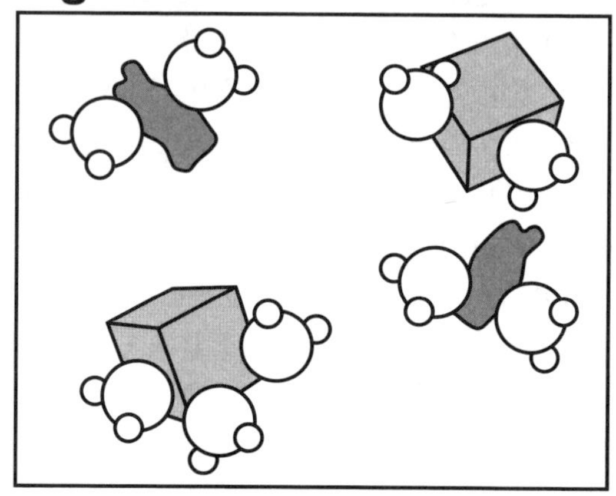

(Not drawn to scale)

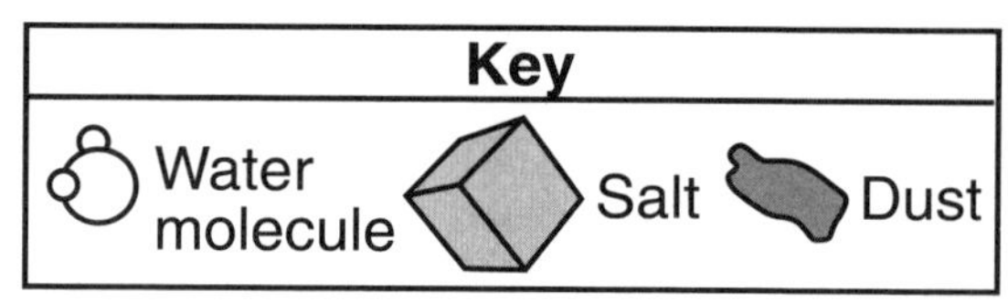

Part 2:

As air rises, it expands. The expanding air cools. When the air drops below the dewpoint, water condenses. This is called adiabatic expansion.

Air generally tracks from west to east across the United States. Sometimes the air, laden with moisture, comes to a mountain. The air mass follows the shape of the land. This event with mountains is called the orographic effect. In the following exercise, complete the directions below using the picture on the next page.

Directions: Complete the following steps using figure 2 on the next page.

1. On the windward side of the mountain, draw the dewpoint line and write "dewpoint"
2. Draw a cloud at the appropriate point.
3. Indicate precipitation in the appropriate place.
4. Draw and label rain.

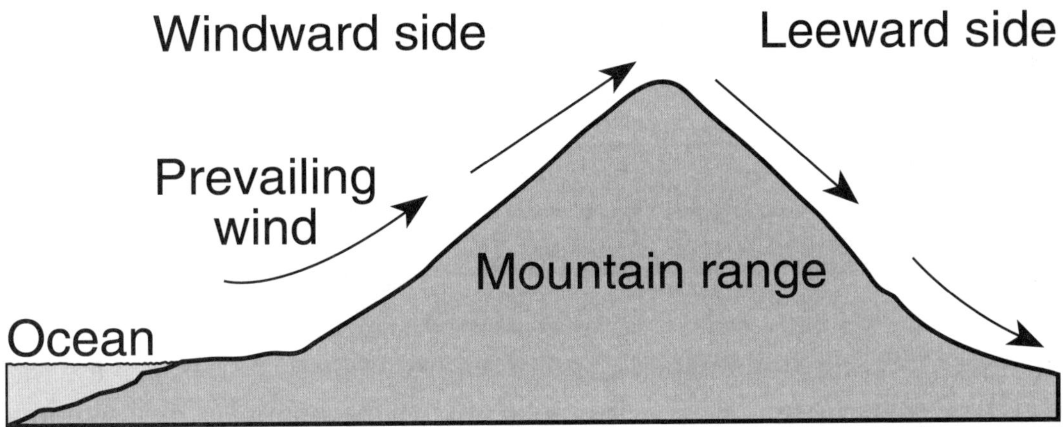

Figure 2

Observations:

1. In relation to clouds, what was the purpose of the smoke in the bottle?

____

____

2. Why was the water swished in the bottle?

____

____

3. Why would you want the bottle to expand?

____

____

4. What does this tell you about cloud formation?

____

____

5. Describe how clouds form.

____

____

____

____

6. Considering the varying temperatures in the atmosphere, why are some clouds composed of ice crystals?

7. Referring to figure 2, notice that the air is descending on the leeward side of the mountain. What do you think is also happening to the air? (Hint: it's the opposite of what happens to air that ascends the mountain.)

8. Referring to figure 2, what happens to clouds as the air mass descends the mountain?

9. Describe what type of climate would be on the windward side of the mountain.

10. Describe what type of climate would be on the leeward side of the mountain.

NAME ____________________

29 Cloud Height

We know that a cloud is a condensed form of atmospheric moisture made up of small water droplets or tiny ice crystals. Because clouds will only form where the water vapor can condense, the clouds must form above the elevation that is below the dew-point. When a cloud falls below that line the water droplets evaporate. This is why clouds often appear to be flat bottomed.

Clouds are categorized by both their height and shapes. High clouds are called cirro and they generally start in height about 6 km above the earth. Middle level clouds have the prefix of alto and they range from 2 km to 6 km. Strato refers to the lowest clouds and they form from the surface to the 2 km level.

Cumulous clouds are puffy clouds that look like cotton. Stratus clouds form sheets of clouds and look like a low grey blanket. Fog is a stratus cloud at ground level. Cirrus clouds are feathery and are formed from ice crystals. Nimbus clouds are dark rain clouds. Nimbostratus clouds have long continuous light rain. Cumulonimbus clouds are large clouds that can extend high above the ground. These clouds are associated with heavy intense rain and thunderstorms. In rare occasions these storm clouds spawn hail or even tornadoes.

Objective: The objective of this lab is to determine the height of the clouds.

Essential Question:

____

____

Hypothesis:

If____

Then____

Because____

Variables:

Independent: ____

Dependent:____

Materials: • Sling psychrometer • Dewpoint chart

Procedure:

1. Using the sling psychrometer, determine the temperature.
2. Using the sling psychrometer, determine the dewpoint.
3. On the chart on the next page, find the temperature (dark, solid lines)
4. On the chart on the next page, find the dewpoint (light, dashed lines)
5. Find where the temperature line and dewpoint line intersect.
6. Altitude is on either side of the chart.

Lapse Rate

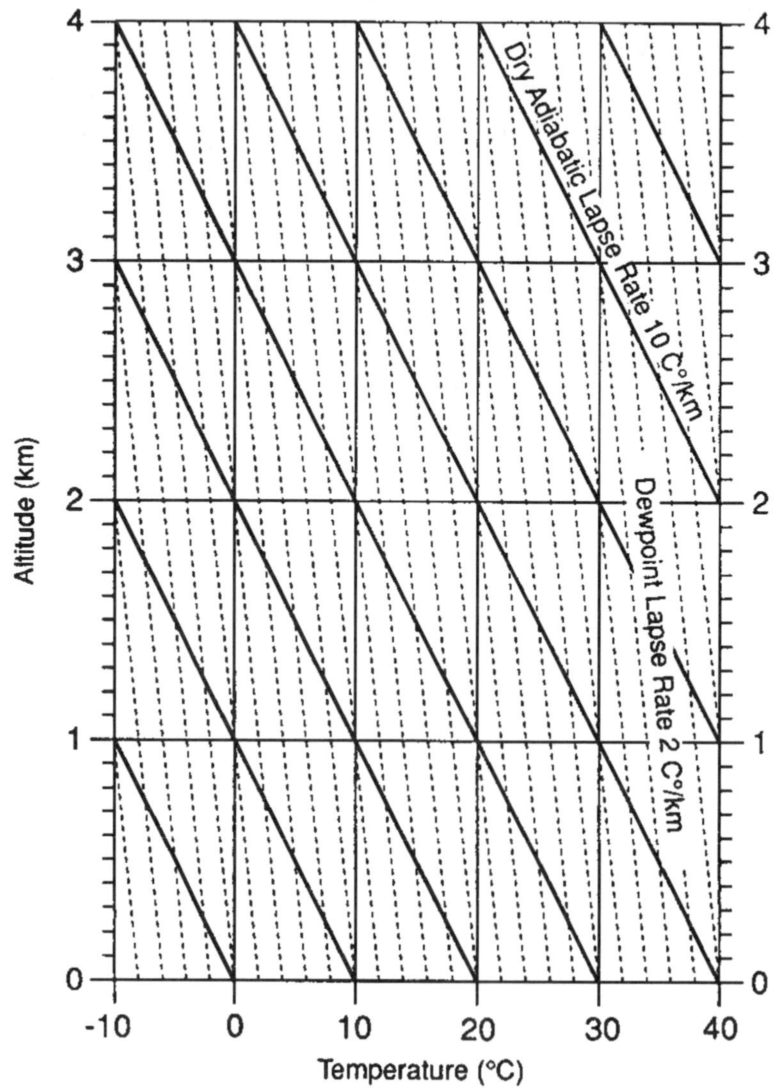

	Dry Bulb (°C)	Wet Bulb (°C)	Dewpoint	Cloud Height
Record Data	°	°	°	km

Dewpoint (°C)

Dry-Bulb Tempera-ture (°C)	Difference Between Wet-Bulb and Dry-Bulb Temperatures (C°)															
	0	1	2	3	4	5	6	7	8	9	10	11	12	13	14	15
−20	−20	−33														
−18	−18	−28														
−16	−16	−24														
−14	−14	−21	−36													
−12	−12	−18	−28													
−10	−10	−14	−22													
−8	−8	−12	−18	−29												
−6	−6	−10	−14	−22												
−4	−4	−7	−12	−17	−29											
−2	−2	−5	−8	−13	−20											
0	0	−3	−6	−9	−15	−24										
2	2	−1	−3	−6	−11	−17										
4	4	1	−1	−4	−7	−11	−19									
6	6	4	1	−1	−4	−7	−13	−21								
8	8	6	3	1	−2	−5	−9	−14								
10	10	8	6	4	1	−2	−5	−9	−14	−28						
12	12	10	8	6	4	1	−2	−5	−9	−16						
14	14	12	11	9	6	4	1	−2	−5	−10	−17					
16	16	14	13	11	9	7	4	1	−1	−6	−10	−17				
18	18	16	15	13	11	9	7	4	2	−2	−5	−10	−19			
20	20	19	17	15	14	12	10	7	4	2	−2	−5	−10	−19		
22	22	21	19	17	16	14	12	10	8	5	3	−1	−5	−10	−19	
24	24	23	21	20	18	16	14	12	10	8	6	2	−1	−5	−10	−18
26	26	25	23	22	20	18	17	15	13	11	9	6	3	0	−4	−9
28	28	27	25	24	22	21	19	17	16	14	11	9	7	4	1	−3
30	30	29	27	26	24	23	21	19	18	16	14	12	10	8	5	1

Observations:

1. How do you find the air temperature using a sling psychrometer?

____

____

____

2. What is the temperature for this day? ______________________________

3. What is the altitude of the clouds? ______________________________

4. Who might need to know the altitude of the clouds? (a profession)

____

____

5. Why do clouds appear to be flat bottomed?

6. What type of clouds do you see today?

7. How do you know it's that type of cloud?

8. What would the elevation be when the humidity is 100%?

9. What do we call those types of clouds in answer 8?

10. How does knowing the height of the clouds help predict what the cloud types are?

NAME ____________________

30 Station Models

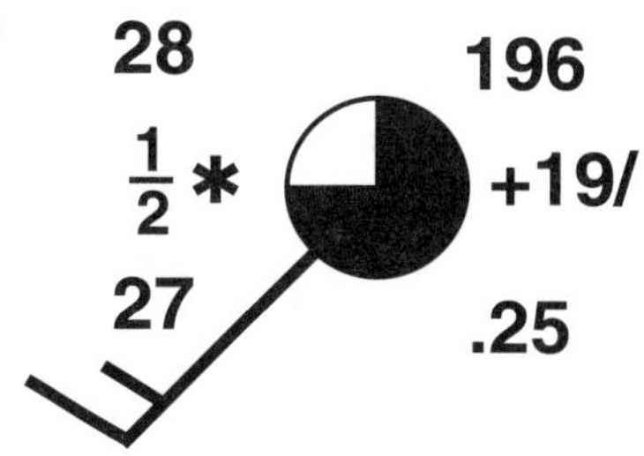

Station models are graphic representations of weather at a particular moment in time. The feathers of a station model always face in a clockwise manner. All of the data that you need to summarize is represented. If you were to look at a regional map, you could find a lot of information in order to make predications. By looking at station models all across a map you can identify fronts, trends and cloud cover. You can also predict the likelihood and the type of precipitation.

In this investigation, you will record weather data and organize it in a station model. The data will be collected over three consecutive days. After the data is retrieved, it will be evaluated.

Objective:

The objective of this lab is to collect weather data and construct a station model and make predictions based on those models.

Essential Question: ____

____

Materials:
- Sling psychrometer
- Directional compass
- Beaufort Wind Scale chart
- Earth Science Reference Tables
- Calculator

Directions: Go outside and follow the procedures below.

Procedure:

1\. Wet the wick of your sling psychrometer.

2\. Using the sling psychrometer find the dew point and humidity.

2\. Estimate wind speed using Beaufort Wind Scale chart.

3\. Find wind direction using a compass.

4\. Record the weather data.

5\. Once weather data is recorded, construct your station model.

6\. Repeat until all station models are done on consecutive days.

Beaufort Wind Scale

Beaufort Number	Description	Characteristics	Range of wind speed knots
1	Light Air	Smoke broken by wind	1 - 3
2	Light Breeze	Wind felt on face	3 - 6
3	Gentle Breeze	Wind extends a light flag	7 - 10
4	Moderate Breeze	Dust and loose paper raise	11 - 15
5	Fresh Breeze	Small, leafy plants begin to sway	16 - 20
6	Strong Breeze	Hard to use umbrellas	21 - 27
7	Moderate Gale	Difficult to walk against the wind	28 - 33
8	Fresh Gale	Twigs broken off tree branches	34 - 40
9	Strong Gale	Chimneys and roofs damage	41 - 47
10	Whole Gale	Uprooting of trees	48 - 55
11	Storm	Damage widespread	56 - 65
12	Hurricane	Extreme damage	Above 66

Weather Map Symbols

Station Model

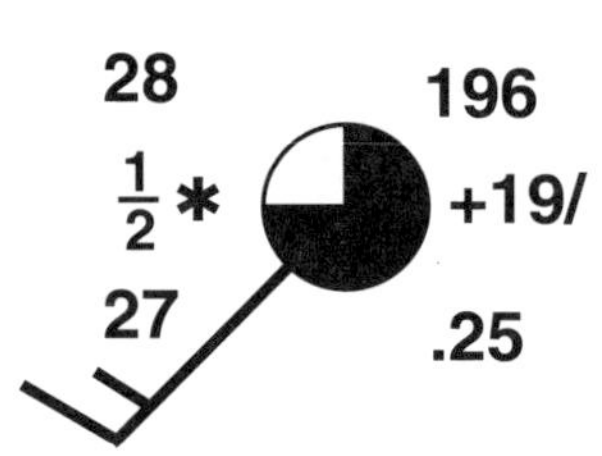

Station Model Explanation

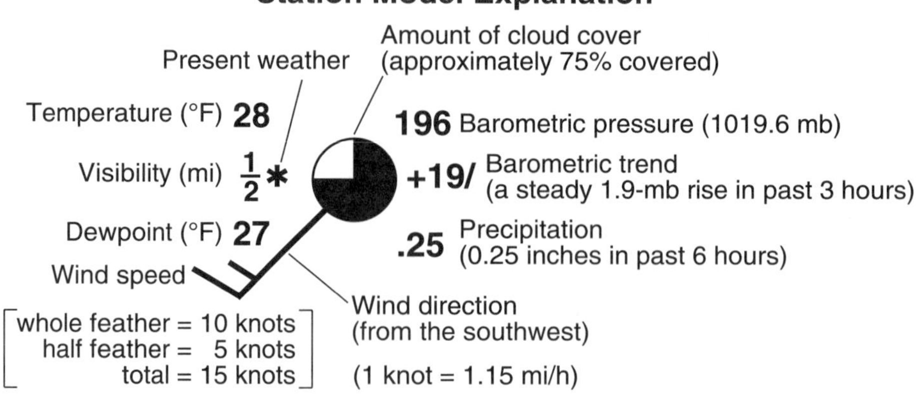

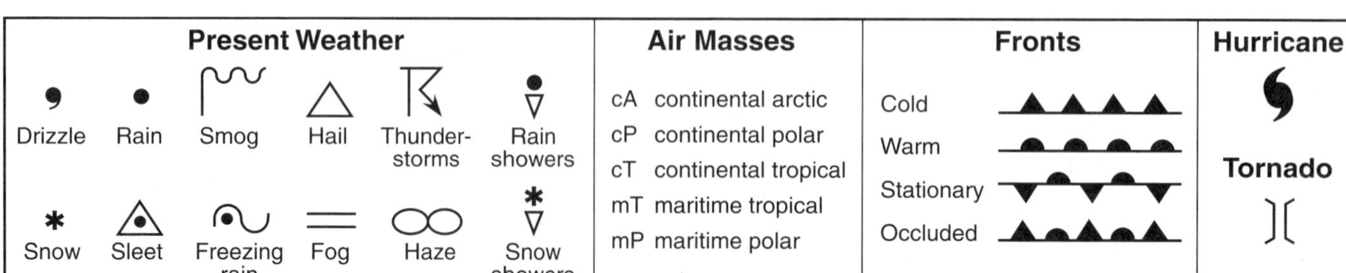

Weather Data

Date__________________

Temperature	Weather Condition	Wind Speed	Dew Point	Cloud Cover	Pressure	Precipitation	Wind Direction
°F			°F		in	in	
°C			°C	%	mb	cm	

What is the relative humidity based on the data retrieved today? ______________________%

Weather Data

Date__________________

Temperature	Weather Condition	Wind Speed	Dew Point	Cloud Cover	Pressure	Precipitation	Wind Direction
°F			°F		in	in	
°C			°C	%	mb	cm	

What is the relative humidity based on the data retrieved today? ______________________%

Weather Data

Date____________________

Temperature	Weather Condition	Wind Speed	Dew Point	Cloud Cover	Pressure	Precipitation	Wind Direction
°F			°F		in	in	
°C			°C	%	mb	cm	

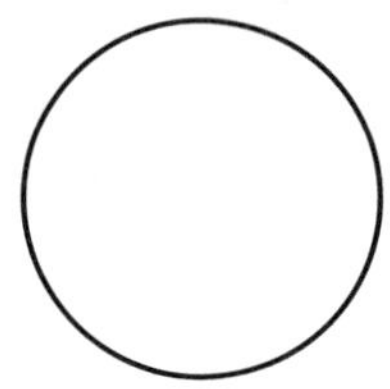

What is the relative humidity based on the data retrieved today? ________________________%

Observations:

1. What changes have you seen in the weather over three days?

____

____

____

2. What is the general trend of the weather?

____

____

____

3. What relationship have you seen between humidity and dewpoint?

4. As the dewpoint is closer to the air temperature, what happens to humidity?

5. What is the relationship of the humidity to cloud cover?

6. How do station models summarize weather data?

7. How do station models help meteorologists predict weather patterns?

Questions 8 and 9 are based on the meteogram below. This meteogram is from John F. Kennedy Airport in New York City on March 10, 2009. Meteograms are ways in which weather changes are recorded for a given area. The meteogram starts at 12:00 am.

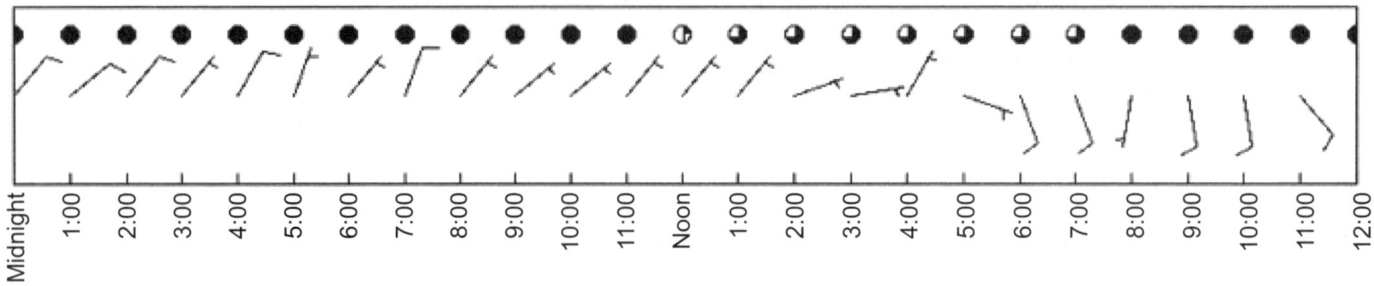

8. Look at the meteogram above. What was the direction of the wind in the morning? Describe the shift of wind for this day.

____

____

____

____

____

____

9. Using the information from the meteogram, calculate the time at which the cloud cover was reduced. ______________________

10. As there was a sudden shift in wind, what happened to cloud cover?

____

____

____

____

____

____

____

31 Weather Maps

NAME ______________________

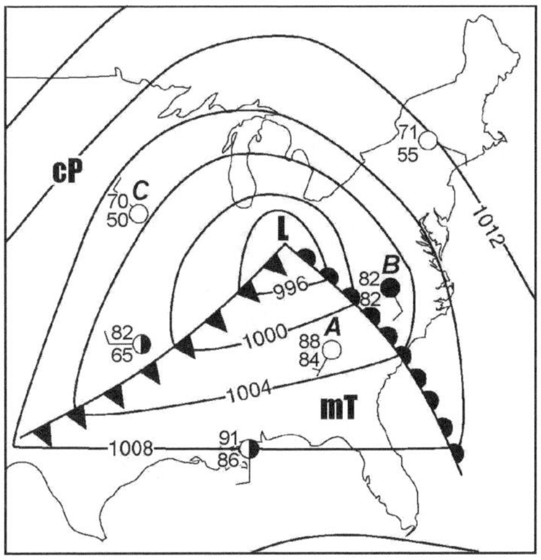

Advances have been made in weather prediction because of computer development, satellites and radar. The World Meteorological Organization, the National Weather Service, and others coordinate weather data collected from around the planet. Data is entered into computer models that apply known information about how atmospheric properties produce forecasts. In some cases, these forecasts have provided warnings of major storms as much as a week in advance. However, because the behavior of weather systems is chaotic, it is difficult to forecast the exact details of weather for more than two weeks in advance.

A weather map shows weather conditions for many places at one time. A key with symbols helps you to read the station models on a weather map. The weather in high-pressure and low-pressure regions is different.

Pre-lab: (done together in class)

- Answer the questions below as part of a class discussion

1. What are two ways to show precipitation on a weather map?

____

____

2. What are areas of low pressure called?

____

3. What is the symbol for high pressure?

____

Objective:

The objective of this lab is to create a weather map based on data in a chart so that one can use to make a prediction.

Materials: • United States map • Colored pencils

Procedure:

Construct a map of isotherms starting with 20° through 60° in intervals of 10°F. Color the field values and correlate to a key of your making.

Temperatures

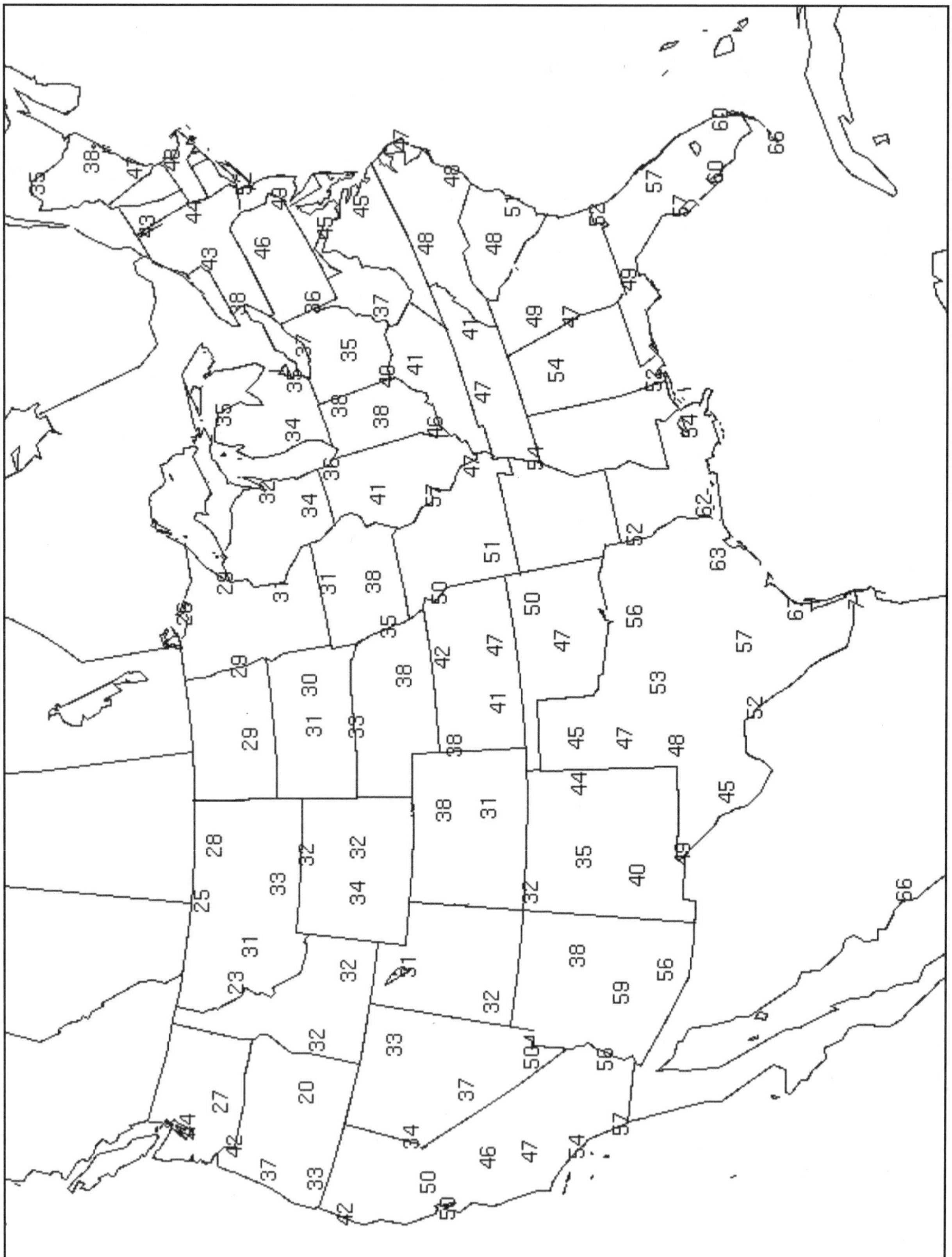

Observations:

1. A ________________________ shows weather conditions for many places at one time.

2. The temperature of a substance is measured by a ______________________.

3. Air pressure always ______________________ with increasing height above the ground.

4. Where do you see higher temperatures, northern or southern United States? Describe the patterns of temperature in the United States.

____

____

____

5. Which has higher overall temperatures, east or western United States?

____

____

6. How could the time of day affect the temperatures?

____

____

7. Describe the temperatures in New York State.

____

____

____

Look at the meteogram for New York City. This weather data was recorded August 10, 2009. Air temperatures are the higher data and dewpoint temperatures are below that.

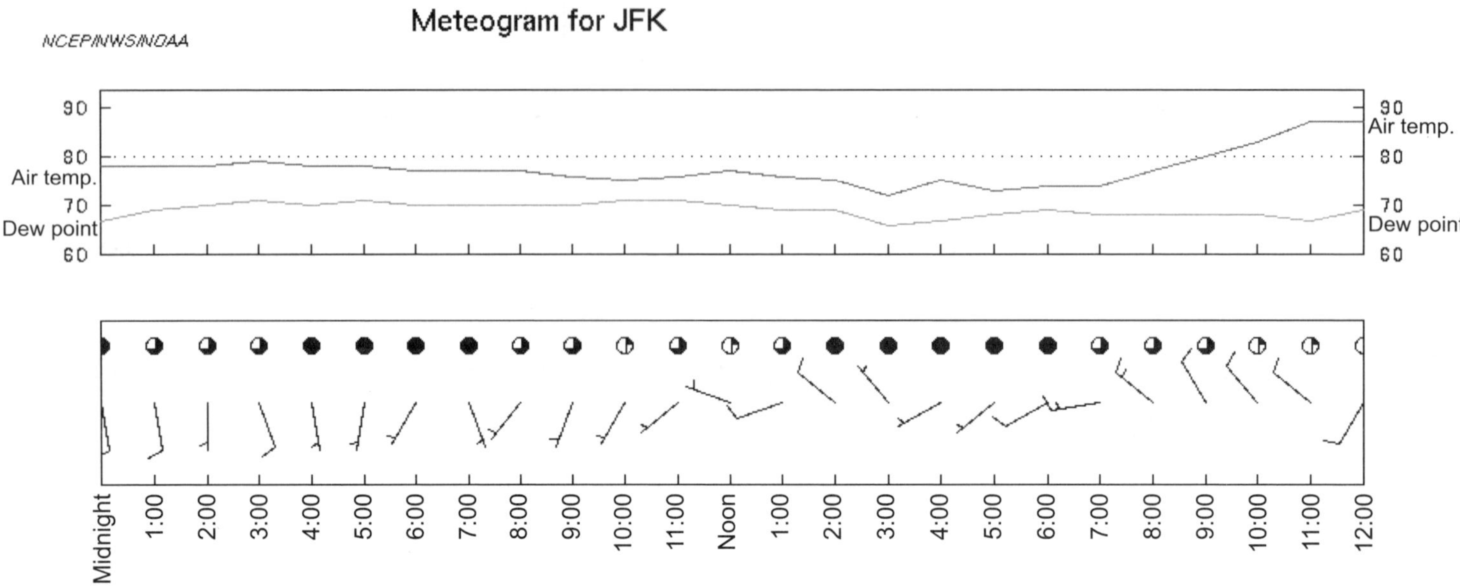

8. What is the wind direction for the morning?

9. When did the wind direction change?

10. In the later afternoon, there was an abrupt change in wind direction. Describe the change in direction and what happened to the temperature and dew point as a result.

NAME ______________________

32 Hurricanes

Hurricanes are the most powerful storms on Earth. They do, in fact, spawn and grow on a global scale. Hurricanes begin as extremely low pressure systems in the eastern part of the Atlantic Ocean. Many form around the Canary Islands off the west coast of Africa. The waters in this area are warmed by the tropical sun. The warm water begins to evaporate. The process of evaporation allows the atmosphere to gain potential energy. The increase of evaporated water in the atmosphere causes the density to decrease. As the pressure decreases, the area is referred to as a Tropical Depression. Tropical Depressions increase in intensity and become upgraded to Tropical Storms. Tropical Storms are larger in area than Tropical Depressions. The winds of a Tropical Storm range from approximately 30 knots to just below 64 knots. Hurricanes are classified as having winds greater than 64 knots. Hurricanes and storms follow along paths that are referred to as steering winds. These are planetary winds, driven by atmospheric properties and the Coriolis effect.

The storm track of Hurricane Floyd started as a Tropical Depression. It increased in strength and became a Tropical Storm and eventually a hurricane. Tracking these changes in wind speed and storm travel are important to study these powerful storms.

Objective:

The objective of this investigation is to determine a pattern to the path and velocity of hurricanes in the Atlantic Basin.

Essential Question:

____

____

Hypothesis:

If____

Then____

Because____

Variables:

Independent: ____

Dependent: ____

Materials:

- Wind Speed Data chart for Hurricane Floyd in September 1999
- Hurricane Tracking chart, Atlantic Basin
- Colored pencils

Date/Time (UTC)	Position		Pressure (mb)	Wind Speed (kt)
	Lat. (° N)	Lon. (° W)		
Sept 8 00:00	**15**	**47**	**1007**	**30**
06:00 am	15	48	1005	35
12:00 noon	16	50	1003	40
06:00 pm	16	51	1000	45
Sept 9 00:00	**17**	**53**	**1000**	**45**
06:00 am	17	54	1003	45
12:00 noon	17	55	1003	50
06:00 pm	18	56	996	60
Sept 10 00:00	**18**	**57**	**995**	**60**
06:00 am	19	58	990	60
12:00 noon	19	59	989	70
06:00 pm	20	60	975	70
Sept 11 00:00	**21**	**60**	**971**	**80**
06:00 am	21	61	963	95
12:00 noon	22	62	962	95
Sept 12 00:00	**23**	**63**	**966**	**90**
06:00 am	23	65	960	95
12:00 noon	23	66	955	105
06:00 pm	23	67	940	115
Sept 13 00:00	**23**	**69**	**931**	**125**
06:00 am	24	70	922	135
12:00 noon	24	71	921	135
06:00 pm	24	73	923	125
Sept 14 00:00	**25**	**74**	**924**	**115**
06:00 am	25	75	927	105
12:00 noon	25	76	930	105
06:00 pm	26	77	930	110
Sept 15 00:00	**27**	**78**	**933**	**115**
06:00 am	28	79	935	110
12:00 noon	29	79	943	100
06:00 pm	31	79	947	95
Sept 16 00:00	**32**	**78**	**950**	**90**
06:00 am	34	78	956	90
12:00 noon	36	77	967	70
06:00 pm	38	75	974	60
Sept 17 00:00	**41**	**74**	**980**	**50**
06:00 am	42	72	983	50
12:00 noon	43	71	984	45
06:00 pm	44	69	985	45
Sept 18 00:00	**45**	**67**	**987**	**40**

Best Track, Hurricane Floyd, September 8 – 18, 1999

Directions:

1. Using the Wind Speed Data for Hurricane Floyd, plot the points on the Hurricane Tracking Chart below by placing an **X** on the chart to represent each point.
2. Circle **X**, in blue the point indicating midnight for each new day and label it.
3. Circle the **X**, in red, at which point Floyd became a hurricane.
4. Circle the **X**, in green, the point(s) at which Floyd's pressure is lowest.

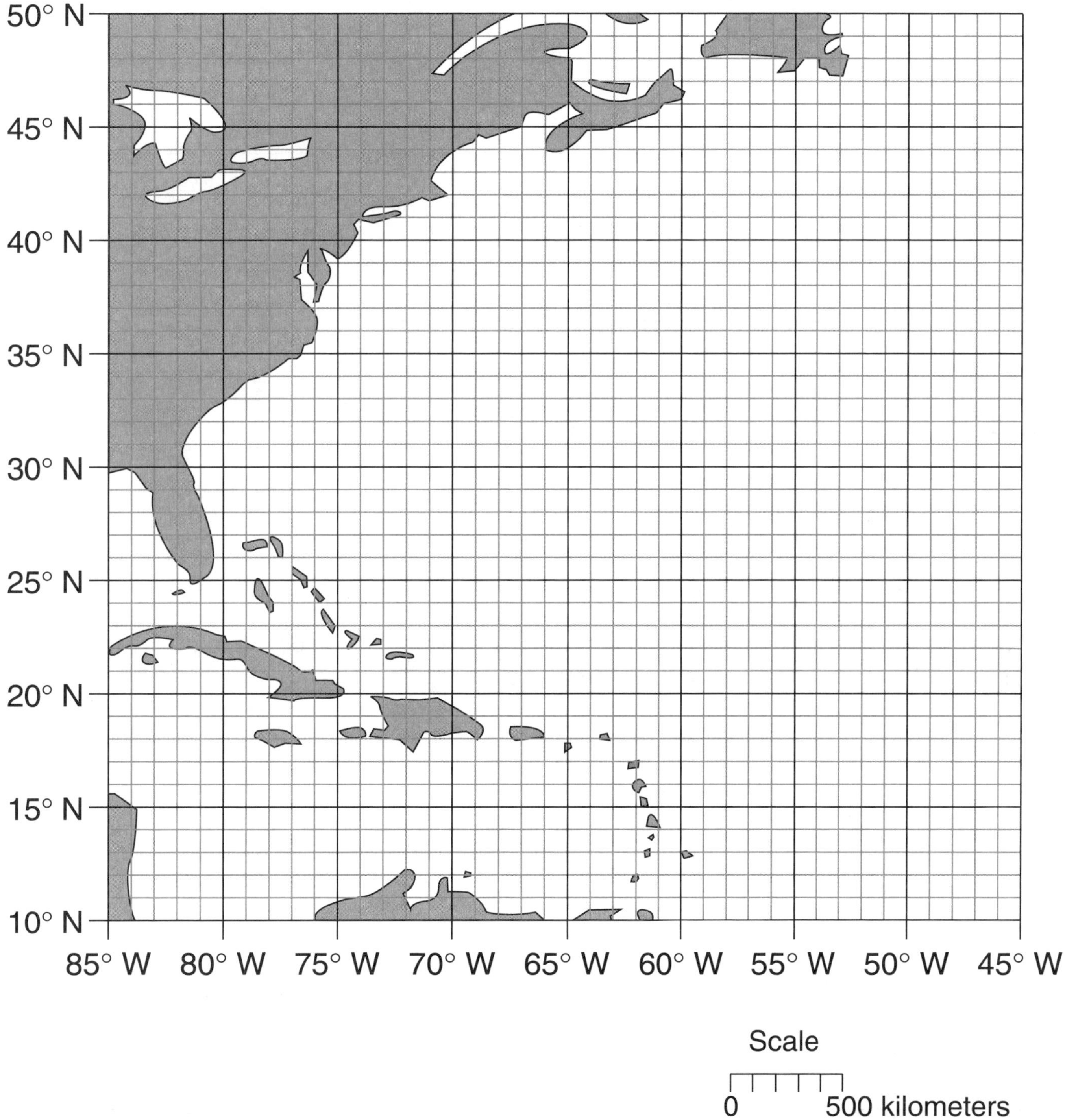

Observations:

1. How much heat does the atmosphere gain as the result of vaporization?

2. What direction did Floyd travel in its initial stage?

3. Did Floyd's sustained wind speed remain constant? Explain.

4. Did Floyd's speed of travel remain constant? Explain.

5. There was a difference in respect to the rate of travel across the ocean from day to night. Explain what may have caused this difference.

6. At approximately which latitude did Floyd change direction?

Use your knowledge and the diagram below to answer questions 7 through 10.

Surface Ocean Currents

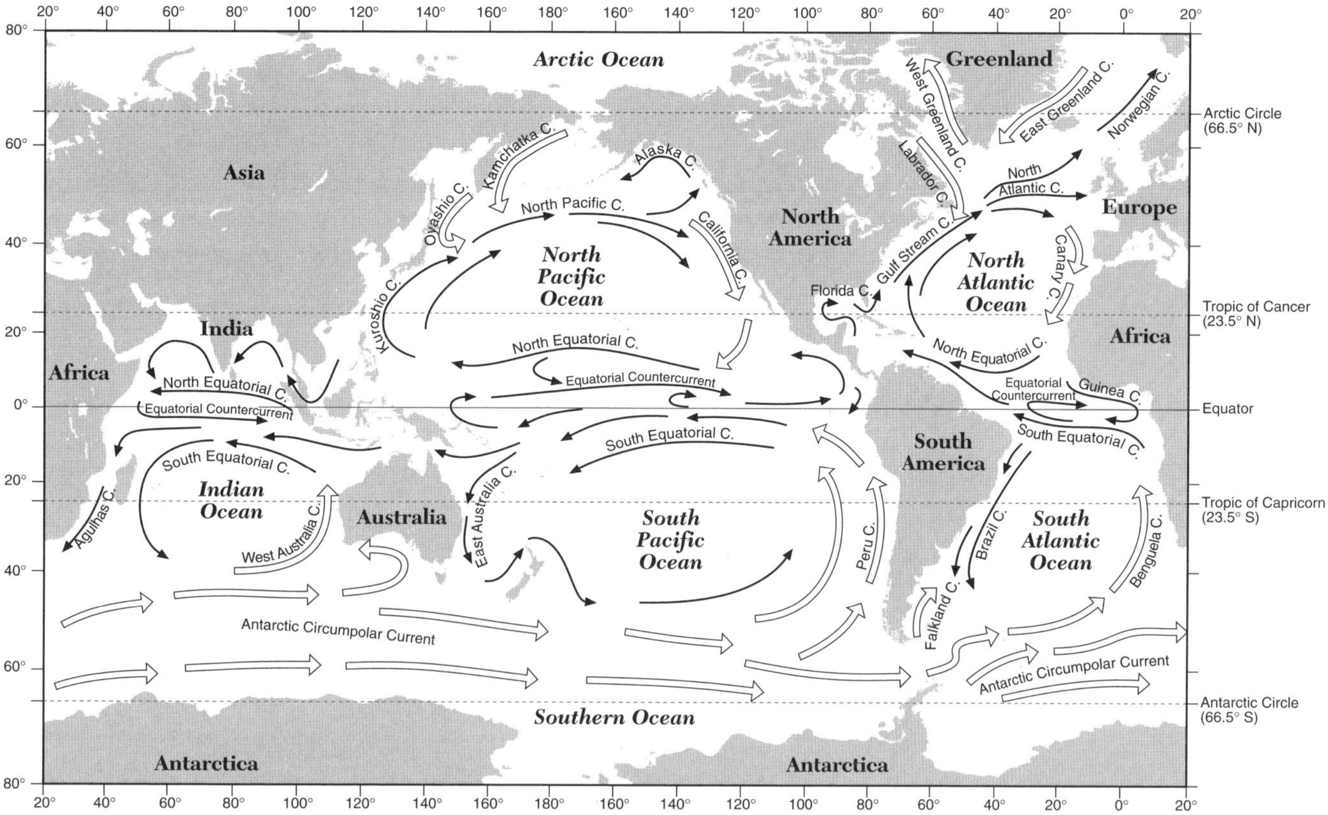

NOTE: Not all surface ocean currents are shown.

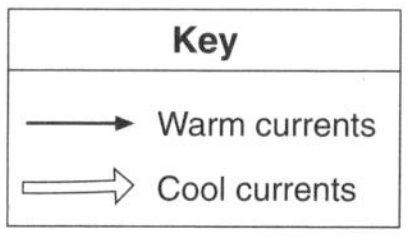

7. Compare Floyd's path as a tropical depression / tropical storm to an ocean current. ____

____

8. Which ocean current did Tropical Storm Floyd travel along?

____

9. As Floyd was upgraded to a hurricane, along what ocean current did Floyd travel?

____

10. Describe how these currents relate to the Planetary Winds.

____

____

____

____

NAME ______________________

33 A Tale of Two Lighthouses

The coastal areas of oceans have features that are created by wave action. The beaches are accumulations of sediment deposited by this wave action.

Long Island is the largest island in the contiguous United States. It lies in southeastern New York State. It is also a very young island. It was formed during glacial period as terminal moraines. Long Island was created in two depositional events. What is now Long Island Sound was a flooded valley when sea levels rose after the melting of the last ice age.

The island is still undergoing changes today. Recall from deposition and erosion, that when depositional forces and erosional forces occur at roughly equal rates, it is called dynamic equilibrium. This means that no one force is greater than the other and so the landform is more or less unchanged. In this investigation we will look at two concerns on the Island. The Montauk Lighthouse and Fire Island Lighthouse were both built over one-hundred years ago. Today, their relation to the ocean is much different then when they were first built.

Objective:

The objective of this investigation is to gain an understanding of coastal processes.

Essential Question:

____

____

Materials: • Coastline chart • Pencil

Pre-lab:

First, let's look at the formation of the coastline. This diagram depicts the Atlantic Coastal Plain.

a. Referring to your Earth Science Reference Tables on page 205, what type of bedrock would you expect to see in this area? ____________________

b. Looking at the area labeled Harbor Hill Moraine, what do think the material in the moraine is made from? ____________________

c. In the outwash plain, what would you expect the particle size to be?

____

d. In the image below of Long Island, there appears to be two areas of deposition that created this island. Looking at the east end, label the Harbor Hill and the Ronkonkoma Moraine.

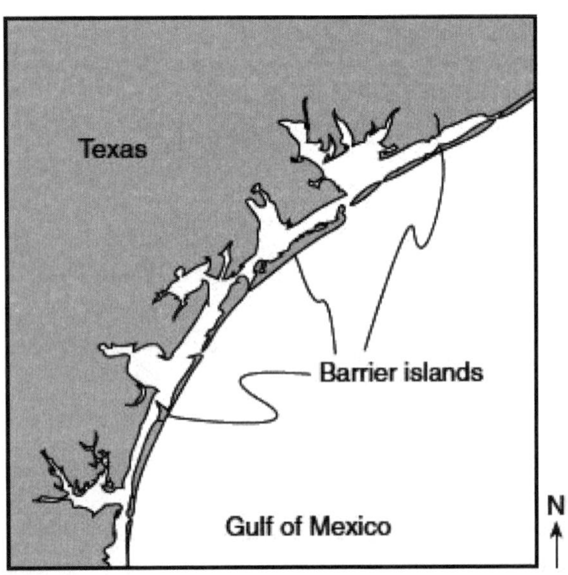

Barrier Islands are islands that form along coastal regions where there is a lot of sand and a passive continental margin. Padre Island is the longest in the United States, over 112 miles long. As seen to the left, barrier islands are long narrow islands that create a protected water area behind them.

New York's Fire Island is also a barrier island and part of the subject of this investigation. The process of erosion and deposition is called littoral drift. In south facing beaches, littoral drift shifts the sand to the west.

Wave Action

Waves are the driving force of the erosional – depositional action of the shoreline. Waves are energy that passes through the ocean. Like other waves, these waves can be refracted and reflected. As waves driven by wind approach the shore, the wave energy touches bottom, creating a frictional force. The refraction of the waves at the shoreline focuses wave energy.

In the accompanying picture, locate the arrows along the bottom of the picture. Above each arrow, draw another arrow up to the next waveform. The arrow must be perpendicular to the waveform. The first row arrows have been done for you. The arrows will reveal a pattern in the waveform. Observe the pattern it creates and answer the questions below.

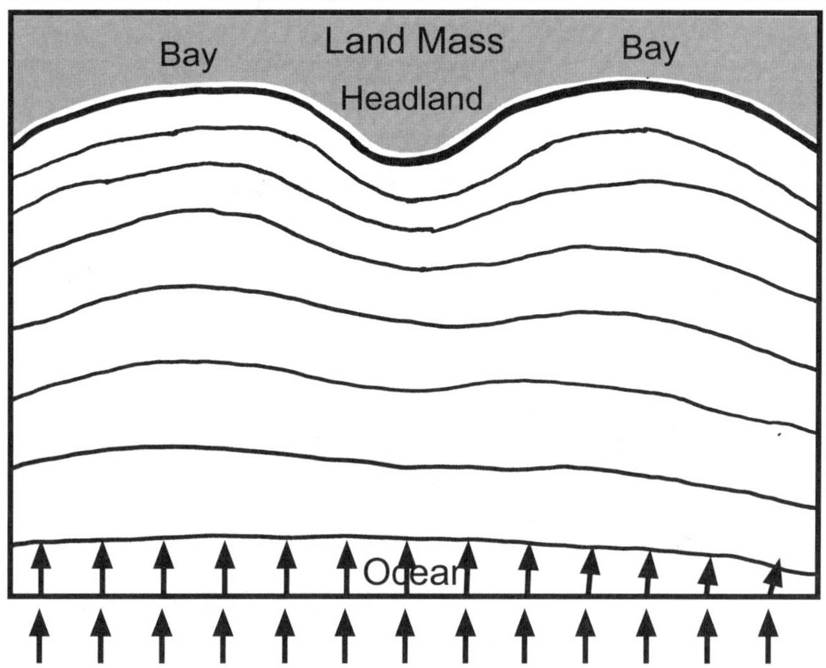

a. Where is the energy focused?

____

b. What do you think is happening at the point(s) of energy focus?

____

____

c. Where did the waves refract from?

____

d. What do you think is happening at the point(s) of refraction?

____

____

e. What will eventually happen to the shoreline?

____

____

Erosion

President George Washington commissioned the Montauk Lighthouse to be built in 1796, The builders knew that Montauk Point was prone to erosion. The original lighthouse was built 300 ft from the sea. 200 years later, the lighthouse stands only 100 feet from the ocean. The lighthouse is in the far eastern end of this point of land. That's 200 feet of erosion in two hundred years. Look at the image below.

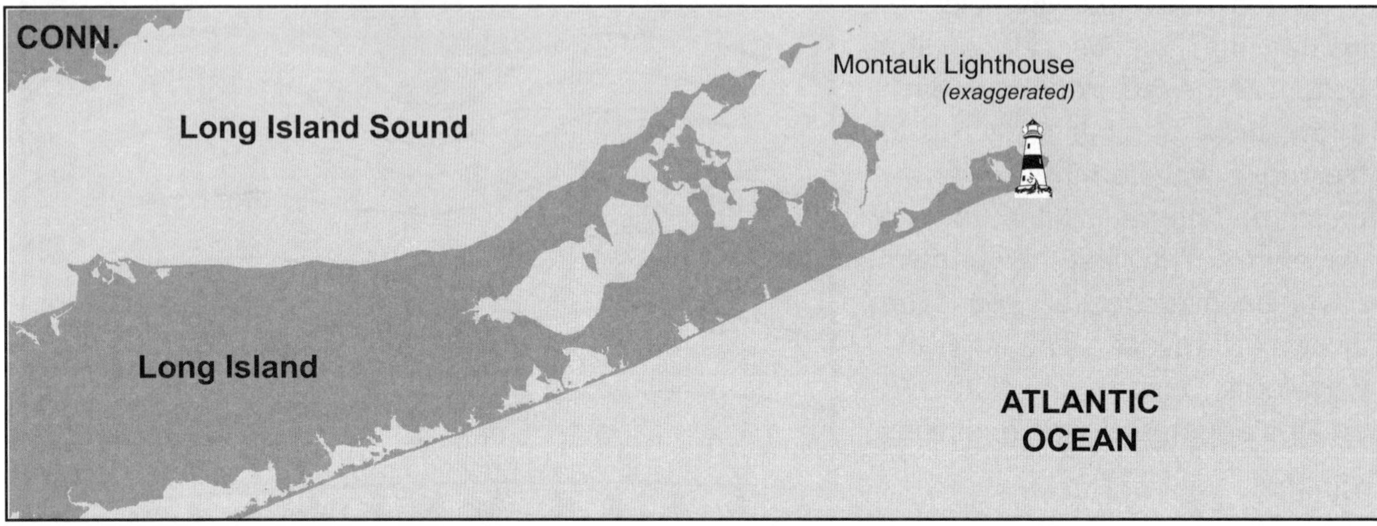

a. Storms are more severe in the winter. Winter storms are called nor'easters. They are in the Atlantic Ocean. Where are the winds of the winter storms blowing from?

____

____

b. Using your knowledge about points of focus, draw an arrow on the image above that most likely would be the point of focus on this land mass.

c. Explain why Montauk is having a problem with erosion.

____

____

d. What would happen if the winter waves were to approach from the southwest?

____

____

e. Give an example of what can be done to slow the erosion.

____

____

Deposition

Waves approach the beach at the angle the wind is blowing them. Sea breezes bring the waves' upwash onto the shore following the waves' momentum. The backwash to the sea is straight back. This ongoing movement is called swash. As waves approach at an angle, the swash is in a zigzag pattern. This moves sediment. The movement that takes place in a motion that moves parallel to the shore is called longshore current. The movement of sediment is called littoral drift.

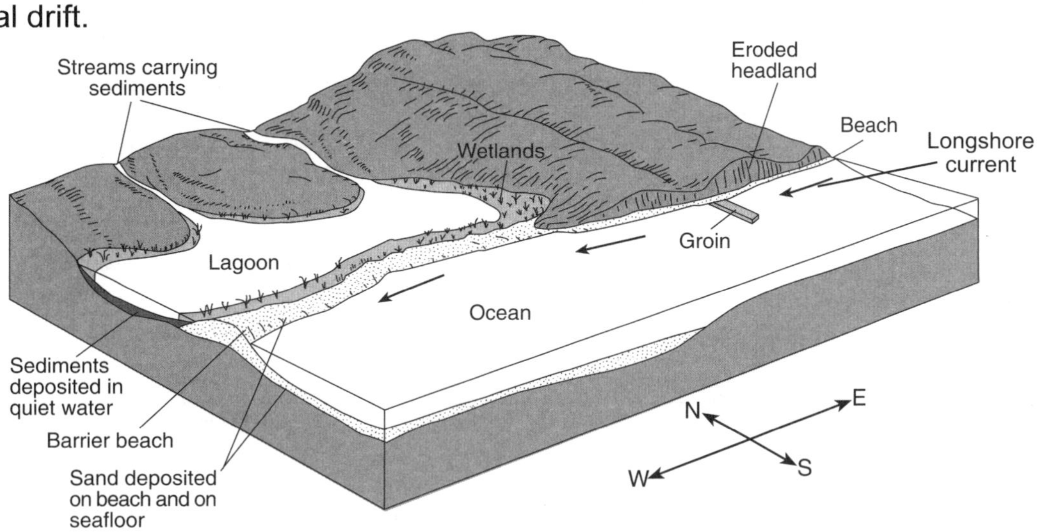

Look at Fire Island. Fire Island is a barrier island to the south of Long Island. It protects the estuary of the Great South Bay. The lighthouse was built in 1858 next to the Fire Island Inlet at the western most section of Fire Island. Today, it is located 6 miles away from the Fire Island Inlet.

When strong winter winds blow from the east, they often take beach with them but in this case, they leave behind winter sand; a reddish purple sand that gets deposited during winter storms. It is made of garnets, magnetite and some quartz. In the summer, quartzite and dolomite materials are left behind.

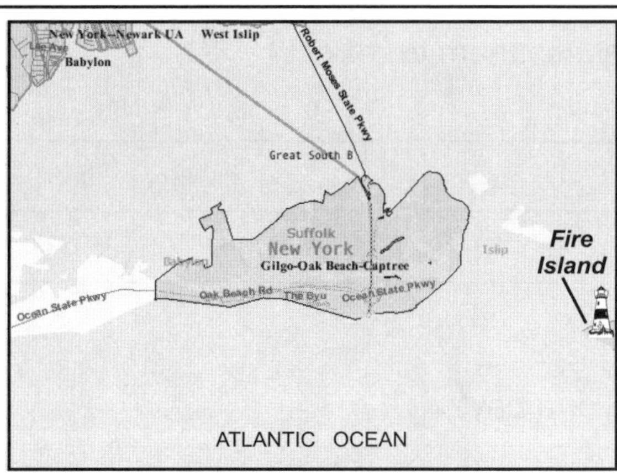

Fire Island 1858

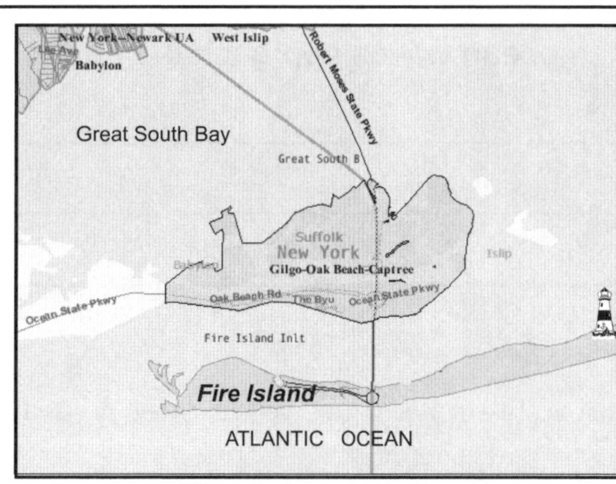

Fire Island Today

a. How fast is the island "moving"?

b. Why is the island "moving"?

c. How do we know that the island has changed shape?

d. What would happen if the winter waves were to approach from the southwest?

e. Give an example of what can be done to slow the deposition.

Observations:

1. When ocean waves approach the beach what causes them to refract?

2. What will receive more wave energy, a headland or a bay?

3. Why will this area receive more energy?

4. If depositional forces were equal with erosional forces, what would we call that?

5. Why is Montauk Lighthouse in danger of falling into the ocean?

6. Which force is more prevalent in Montauk, erosion or deposition? Why?

7. Where does the winter sand at Fire Island come from?

8. If unchecked, what will happen to Fire Island Inlet?

9. Looking at the bedrock of Long Island and the Atlantic Coastal Plain, explain the role of erosional and depositional forces that created and are shaping the island.

10. Describe what changes may happen to this region of New York State.

NAME ____________________

34 Seasons

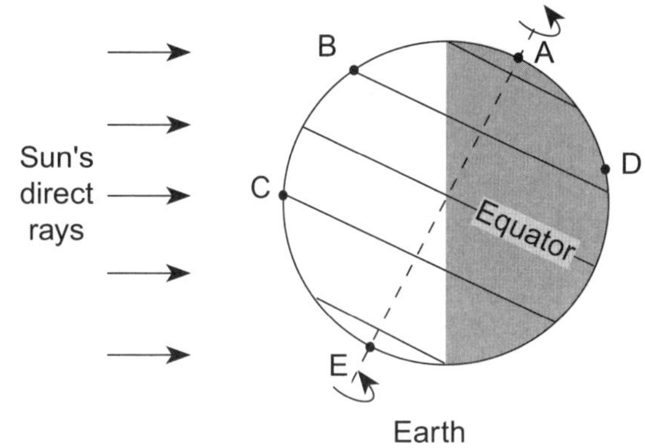

The Earth axis is tilted at 23.5° relative to the Sun. It is because of this tilt that we have seasons. The Vernal Equinox marks the beginning of spring in the Northern Hemisphere; the Sun passes directly above the equator, or 0°. This generally occurs around March 21. During the Summer Solstice, summer begins in the Northern Hemisphere. The Sun passes over the Tropic of Cancer, 23.5° N. It occurs around June 21. The Autumnal Equinox, or the first day of fall, usually happens around September 21. As the Sun makes its apparent trip south, it passes over the equator again. The Winter Solstice occurs or on December 21. When the Sun's direct ray reaches the Tropic of Capricorn, 23.5° S; winter begins in the Northern Hemisphere. In this lab, we will be looking at the Earth – Sun relationship from both a heliocentric view and a geocentric view.

Objective:

The objective of this lab is to demonstrate an understanding how the Earth – Sun relationship affects the seasons.

Materials:

- Diagram of the Celestial Sphere
- Earth Science Reference Tables
- Clear plastic hemisphere
- Protractor
- China pen or dry erase marker

Procedure:

1. Complete the Heliocentric Season chart on the next page using the information in the text.
2. Fill in the name of the season. (winter, summer, etc.)
3. Fill in the first day of the season, like June 21.
4. Indicate where the Sun's vertical ray is on that first day of that season.
5. Give the name of the latitude. Some latitudes have names like the Arctic Circle.

Season Chart

	Season	Date of First Day of Season (Approximate)	Latitude of Vertical Solar Ray	Name of Latitude
Vernal Equinox			°	
Summer Solstice			°	
Autumnal Equinox			°	
Winter Solstice			°	

So, if we were to look at the Sun's path, it might appear like this:

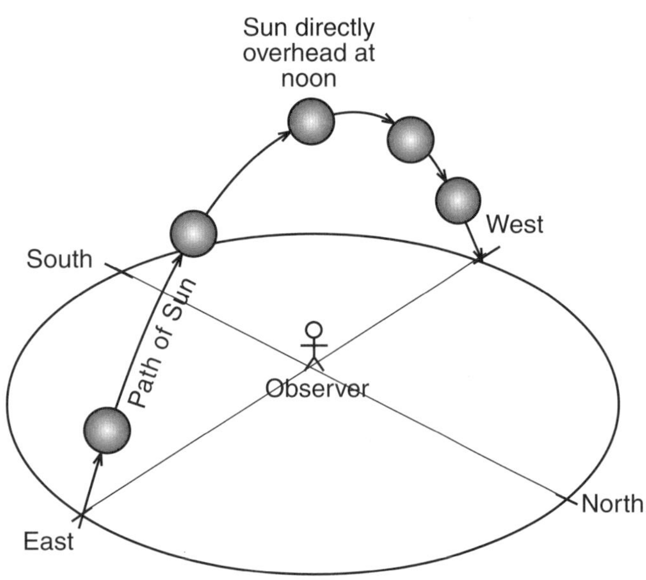

What is the name of the latitude that the Sun's vertical ray is over in the picture?

____

What might possibly be the date in the picture to the left?

____

The picture above shows the Sun's path arching across the sky or a celestial hemisphere.

We know that the Earth – Sun tilt is at an angle of 23.5°. So the Sun's vertical rays appear 23.5° N of the equator on June 21, causing the first day of summer. Similarly, the Sun's vertical rays appear in the Southern Hemisphere at 23.5° S.

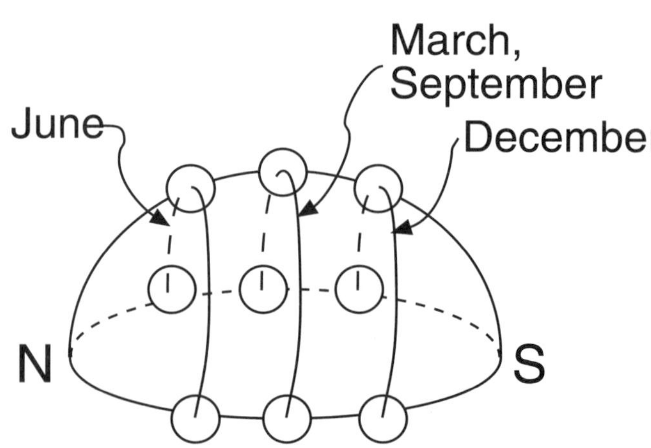

Sun's Path at the Equator

In the space below, calculate the angles of insolation for the following days.

Summer Solstice:

______________________________ + 23.5 = ______________________________

angle of insolation for equinox   angle of insolation for summer solstice

______________________________ – 23.5 = ______________________________

angle of insolation for equinox   angle of insolation for winter solstice

We get an angle of insolation of ______________°. In New York, the Sun is never overhead in the northern sky. Place an **X** where the Sun's angle is during the equinox. The Sun rises due east and sets due west during the equinox. As in the illustration on the previous page, draw the Sun's path on the equinox. IMPORTANT: Keep in mind where the north is.

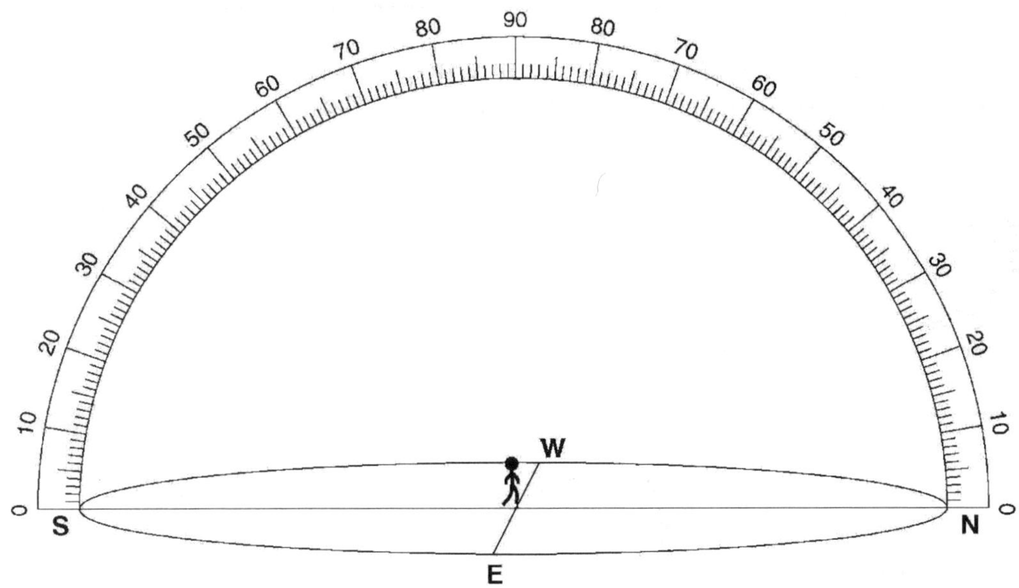

We've established an angle of insolation for the equinox. Now you will find the angle of insolation for summer solstice and winter solstice.

The formulas are as follows:

Formula for Calculating the Angle of Insolation

Angle of insolation for the equinox	+ (plus)	23.5°	=	Summer solstice
°	+	°	=	°

Angle of insolation for the equinox	– (minus)	23.5°	=	Winter solstice
°	–	°	=	°

Procedure:

1. Place a clear plastic hemisphere on the compass rose on page 172.
2. Locate "South" on the compass rose.
3. Line the external protractor at South up to the top of the hemisphere. 0° should be at the equator and 90° at the top.
4. Use the external protractor to located the winter solstice, the equinoxes and the summer solstice. Mark these points on your hemisphere using the china pen. These points represent Noon.
5. Find East on the compass rose. Connect the points from due East to Noon for the equinox you made in procedure 3. Find West. Connect the points from Noon to due West.
6. Making parallel lines to the equinox path, mark the Sun's path for the summer solstice "rising in the east" passing through Noon, and "setting in the West".
7. Repeat Step 6 for the winter solstice.
8. Knowing how far the sun appears to travel in one hour, mark off each hour from Noon. This should be done for all three of the Sun's paths.

Observations:

1. Why are there different paths for the different times of the year?

____

____

2. Which day is represented by the shortest path?

____

____

3. Looking at the paths for the equinox, why is there only one line?

____

____

4. Looking at the paths on your hemispheres, was there a significant difference in distance of the paths traveled? Describe.

____

____

5. Equinox literally means equal night. What information that you tracked on your hemisphere would support this statement?

____

____

6. What is the time it takes for the Sun to track during the equinox?

____

7. Which day has the greatest duration of sunlight? Explain.

____

____

____

8. Look at the Sun's path for the winter solstice.
In what direction does the Sun rise? ______________________

9. Look at the Sun's path for the summer solstice.
In what direction does the Sun rise? ______________________

10. Describe the height of the Sun at 6:00 pm on December 21 in the Northern Hemisphere.

____

____

____

____

____

____

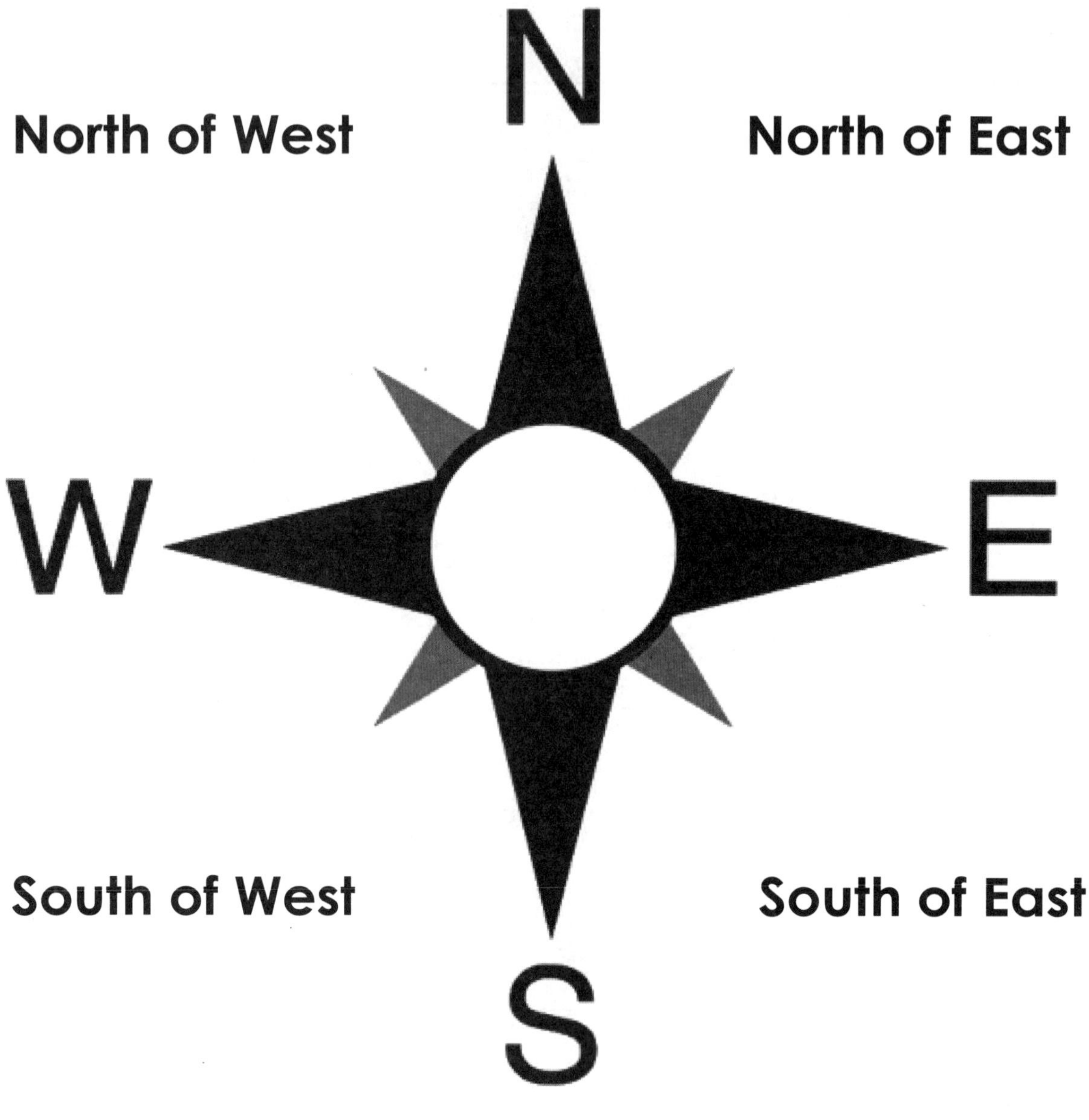
N
North of West
North of East
W
E
South of West
South of East
S

NAME ____________________

35 Electromagnetic Spectrum

Light is a form of energy visible to the human eye that is radiated by moving charged particles. Scientists have learned, through experimentation, that light sometimes behaves like a particle and at other times like a wave. The particle features of light are called photons. When light refracts, or bends slightly as it passes around a corner, it shows wavelike behavior. These waves associated with light are called electromagnetic waves. They consist of changing electric and magnetic fields. Light has many qualities, including energy and wavelength that cannot be seen by human eyes.

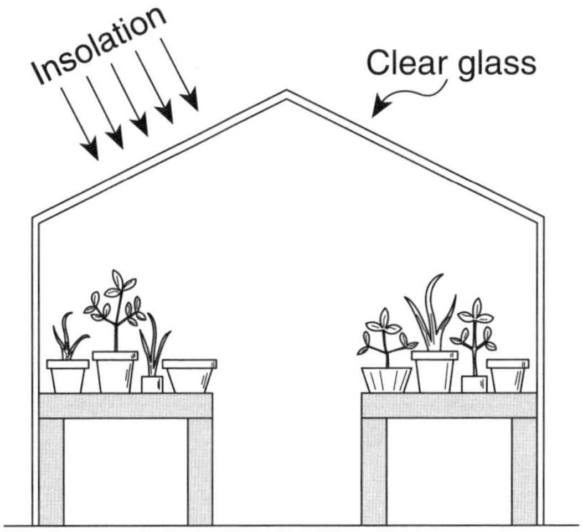

Greenhouse

Objective:

The objective of this lab is to gain an understanding of the properties of the electromagnetic spectrum.

Essential Question:

____

____?

Hypothesis:

If ____

Then ____

Because ____

Materials:

- Spectroscope
- Ultraviolet "black" light
- Fluorescent material
- Shadow box
- Remote control
- Thermometer
- 5 cm square white reflective material
- 5 cm square white rough material
- 5 cm square black reflective material
- 5 cm square black rough material
- Slinky
- Earth Science Reference Tables

Procedure:

Follow the directions listed for each station. With your lab group, go to each station and perform the activity as outlined. You do not need to go in order, but complete all stations. If a station is full, fill in "Observations".

Under no circumstances are you to look directly into any light source!

Electromagnetic Spectrum

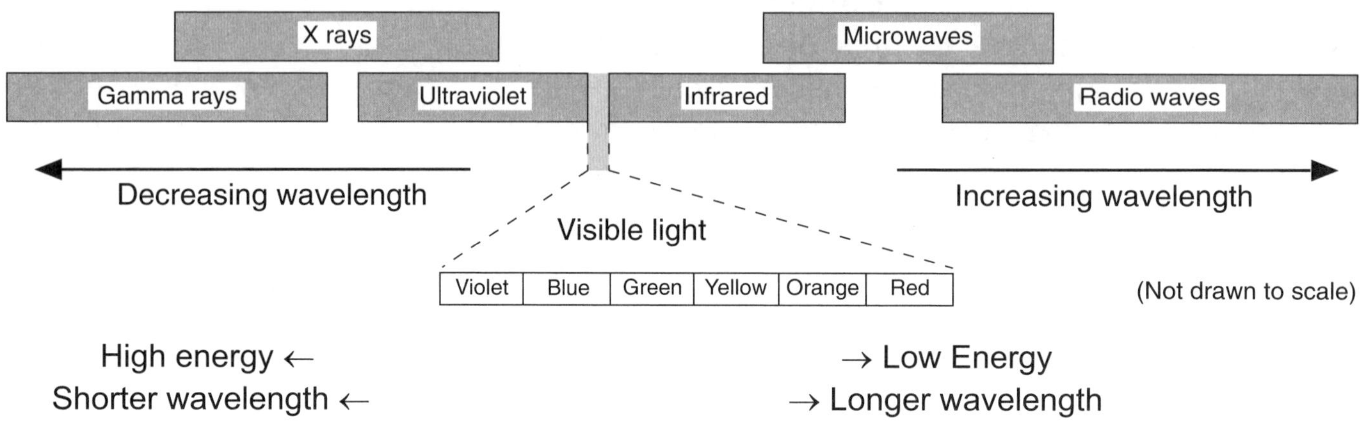

Station One: Using the spectroscope, observe different light sources. Using your colored pencils, color in the space provided the colors you observed.

	Colors observed (indicate if you cannot see all colors)
1. Sunlight	
2. Fluorescent light	

Observe:

Compare and contrast the two light sources that you observed.

Infer:

Based on your knowledge of the electromagnetic spectrum, what is the cause of these differences, if any?

Station Two: Place the string of beads under the ultraviolet light in the light box.

Observe:

What do you see with ultraviolet light that you can't see with visible light?

Infer:

Based on your knowledge of the electromagnetic spectrum, what is the cause of these differences, if any?

Station Three: Remotes

We know remotes transmit energy that cause the electronic device it's linked to, to turn on or off. You and your partner examine the remote at this station. Look for the light emitted. Press any button, that will cause the remotes to give off light. Describe what you see.

Repeat what you did above. Now use the camera your teacher described.
Describe what you see.

Station Four: Place the slinky at the ruler on the back of this book. Grasp both ends so that a small portion of the slinky is on the ruled edge, with your left hand aligned to the beginning of the cm line. Have the coils of the springs spread so that each coil is 1 cm apart

Your right hand should be on 10 cm mark.

1. Pull your right hand to the 25 cm mark.
   a. Did your right hand go toward or away from your left hand? ___________________
   b. Record any changes in the wavelength.
   ____
   c. If your original wavelength represented visible light, what portion of the visible spectrum would the color shift to?
   ____

Your right hand should be on 10 cm mark.

2. Push your right hand to the 5 cm mark.
   a. Did your right hand go toward or away from your left hand? ___________________
   b. Record any changes in the wavelength.
   ____
   c. If your original wavelength represented visible light, what portion of the visible spectrum would the color shift to?
   ____

Station Five: Reflection

Look at the four surfaces.

1. Take a thermometer from the supply tray and record the temperature reading before using.
2. Place a thermometer under each of the four square samples of material.
3. Turn on the heat lamp.
4. After 2 minutes remove from heat and record data.

	Dark Rough Sample	Dark Reflective Sample	Light Rough Sample	Light Reflective Sample
Before				
After				

Infer:

Based on your observations, does color make a difference in energy absorption?

____

____

Based on your observations, does texture make a difference in energy absorption?

____

____

Station Six: Rayleigh Scattering

Sunlight hits our atmosphere and is both reflected and absorbed. The longer wavelengths of visible light pass through the atmosphere easily. The shorter wavelengths get absorbed by the gas molecules in the atmosphere. These molecules in turn radiate light at approximately a wavelength of about 4.3×10^{-5} to a wavelength of 4.9×10^{-5}. Since the atmosphere is emitting this light the color of the sky looks the same, although paler along the horizon.

Place the spring at the ruler on the back of this book. Grasp both ends so that a small portion of the spring is on the ruled edge, with your hand aligned to the beginning of the cm line. Have the coils of the springs spread so that each coil is 1 cm apart.

1. Look at the back page of the lab book. What is the color of light emitted by the atmosphere?

____

Your right hand should be on 5 cm mark. Gradually pull your right hand so that the wavelengths become longer. Keep in mind as the sun sets, longer wavelengths pass through greater amounts of atmosphere.

2. If we know that the longer wavelengths of the visible spectrum pass readily through the atmosphere, what are the colors that get scattered in the atmosphere?

____

3. What is the scattering called?

____

4. Why is the sky blue?

____

____

Directions: Use the Electromagnetic Spectrum chart below to answer questions 1-3.

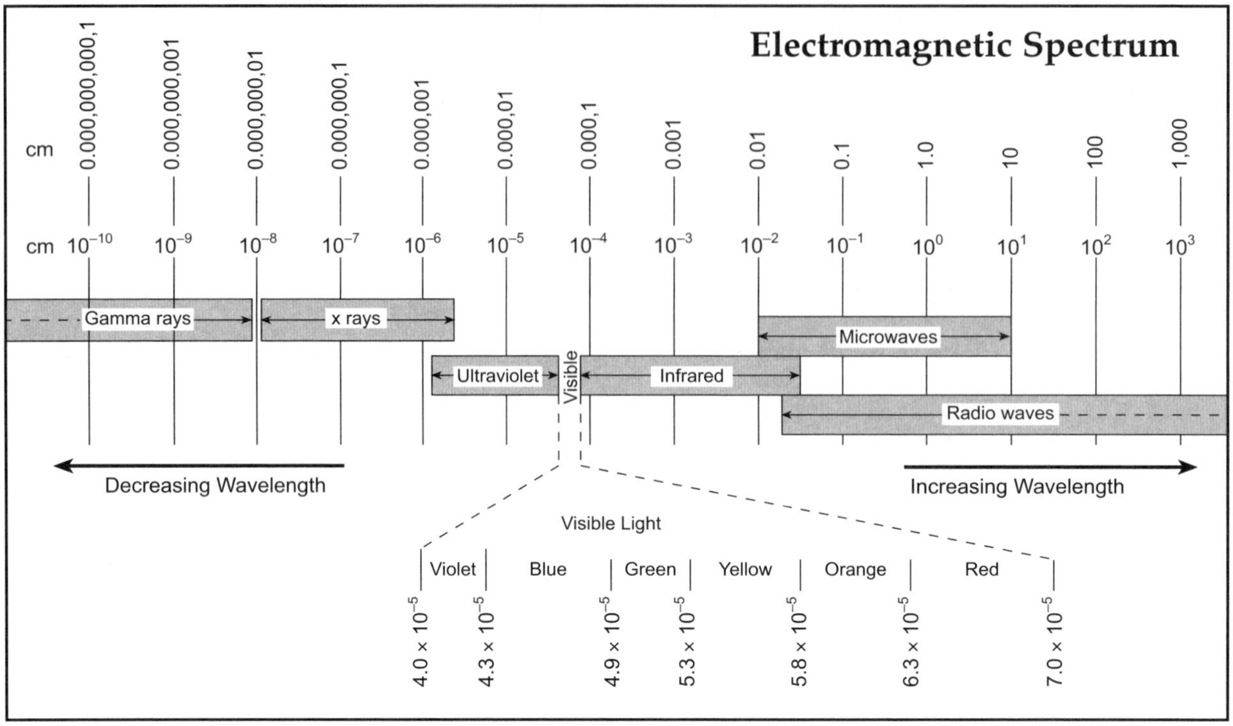

Observations:

1. What is the fundamental difference of the segments of the various light groups found on the Electromagnetic Spectrum chart?

____

2. What is the approximate range of wavelengths of visible light?

____

3. Which has a longer wavelength, ultraviolet rays or infrared rays?

____

4. Referring to Station One, if we split sunlight and fluorescent light, and saw different light, how are rainbows made?

____

____

5. Referring to Station Four, question 1, what do you think happens to the light emitted when astronomers see stars moving away from them?

____

____

____

6. Why do you think people would want something that detects ultraviolet light?

____

____

____

7. Referring to Station Four, question 2, what do you think happens to the light emitted when astronomers see stars moving toward them?

____

____

____

8. Referring to Station Five, describe the surface that was a better absorber of light energy.

____

____

____

9. Referring to Station Five, describe the surface that was a better reflector of light energy.

____

____

____

10. Assuming your lab partner was absent today, write a brief note to your partner summarizing new information you learned today about the properties of the Electromagnetic Spectrum chart.

NAME ______________________

36 Where is Our Sun?

Ancient people believed that the Sun and all of the celestial objects revolved around the Earth. This is known as the geocentric model. Ancient people closely watched the movement of the Sun. These people built simple devices called sundials, to tell time. The sundial works on the notion that the sun moves at regular speed across the sky. A gnomon is the shadow cast by the sun on a sundial. Knowing the angle of the Sun and your longitude, one can easily determine the local time. Local time is when people determine noon to be when the Sun appears to be at its highest point in the sky. When people need to determine true north, people need to know the difference between where the compass is pointing and the North Pole. This difference is called magnetic declination.

Objective:

The objective of this lab is to determine the angle of the Sun and the local time.

Essential Question:

____

____

Hypothesis:

If ____

Then ____

Because ____

Variables:

Independent: ____

Dependent: ____

Materials:

- Two meter sticks
- USGS Quadrangle
- Compass (directional)
- Protractor

Procedure:

1. Find the magnetic declination on the USGS Quadrangle.
2. Go to an area of direct sunlight (outside)
3. Find a level place of ground.
4. Have one partner hold one end of the meter stick loosely until it touches the ground. This will allow the stick to be perpendicular to the ground.
5. Finding the shadow it casts, place the second meter stick on the ground, along its gnomon. Be sure to place the lower numbers of the meter stick at the vertical meter stick.
6. Place the compass on the ground, next to the meter stick. Make sure your angle is less than 90°. Adjust the dial so that while the compass is still on the ground, the arrow aligns with the needle. This is the arc of the shadow.
7. Measure the gnomon. (Read the line where the shadow falls. Zero is facing the vertical meter stick.)
8. Draw your shadows (scale 1:10) in the space below. We know that the vertical line of the right triangle is 1 meter and the gnomon is the horizontal length.

TIME CHART

Magnetic Declination	Arc of Shadow and magnetic declination	Rotation / hour	What time is it?
	÷		

Remember, there are 60 minutes in an hour, adjust your decimal from division accordingly. If it is morning, subtract hours and minutes from noon. If it is afternoon, add hours and minutes to noon.

Observations:

1. Define magnetic declination.

____

____

2. What is the local magnetic declination?

____

3. Define gnomon.

____

____

4. What is the length of the gnomon? ______________________cm.

5. What is the angle of the Sun? ______________________°

6. How many degrees does the sun appear to travel in an hour? ______________°

7. What is the difference between magnetic north and the angle of the gnomon? (The directional compass will show you.)

____

____

____

____

8. What is the local time? ______________________________

9. Why is the local time so different than standard time?

____

____

____

____

____

____

10. If we know the time in Greenwich, England, and we know our local time, what else can we find out?

____

____

____

____

____

____

NAME ______________________

37 Lunar Orbit & Phases

An observer can see only 50% of the Moon's entire surface. However, an additional 9% can be seen from time to time around the apparent edge because of the relative motion called libration. This is because of the slightly different angles of view from Earth, due to different relative positions of the Moon along its inclined elliptical orbit.

One revolution of the Moon around Earth takes a little over 27 days 7 hours. This is called a sidereal month. The Moon rotates on its axis in this same period of time, so the same face of the Moon is always presented to Earth.

Over a period, a little longer than 29 ½ days, the Moon goes through phases, in which the lit half of the Moon that we see from Earth changes. This is called a synodic month. The phases are caused by the changing angle of sunlight hitting the Moon. The time of phases is longer than the period of revolution of the Moon, because the motion of Earth around the Sun changes the angle at which the Sun's light hits the Moon from night to night.

Essential Question:

____

____

Objective:

The objective of this lab is to determine lunar phase relative to Moon's position in orbit.

Materials:
- Moon's Position in Orbit chart
- Lunar Phases chart
- Lunar calender

Procedure:

1. In Figure 1, with a pencil, extend the arrows between moon phases to make a complete circle. Note that this line represents the part of the Moon visible from the earth.
2. Use a pencil to shade in the dark side of the Moon at each of the eight moon positions.
3. In Figure 2, show how the moon would look from the Earth at each position, by shading in the dark part. Number 7 has been done for you.
4. Label each of the phases of the Moon in Figure 2: *Full, Gibbous, Quarter, Crescent or New.*

Figure 1

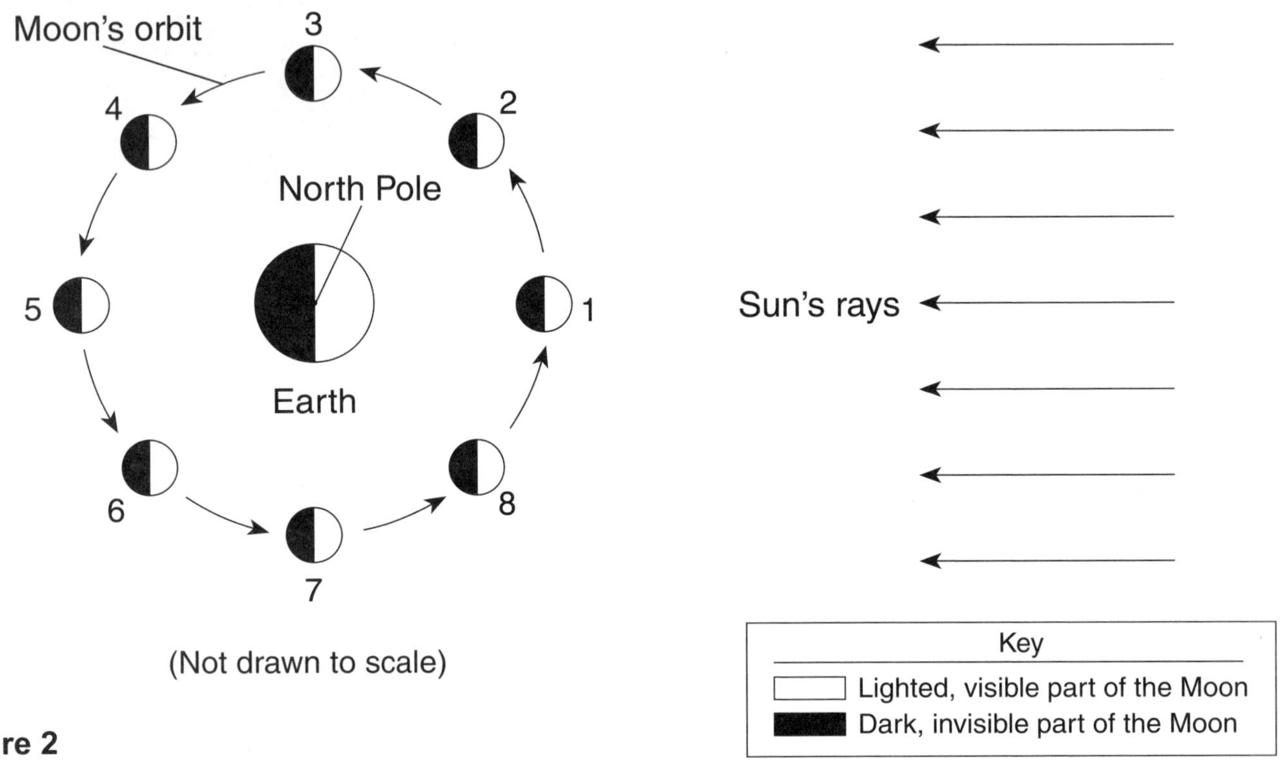

(Not drawn to scale)

Figure 2

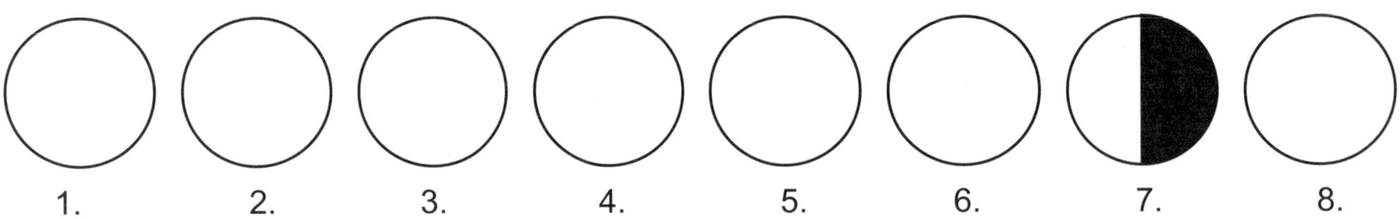

1. ______________________ 5. ______________________

2. ______________________ 6. ______________________

3. ______________________ 7. *Quarter*

4. ______________________ 8. ______________________

July

Sunday	Monday	Tuesday	Wednesday	Thursday	Friday	Saturday
1	2	3	4	5	6	7
8	9	10	11	12	13	14
15	16	17	18	19	20	21
22	23	24	25	26	27	28
29	30	31				

Key

- New Moon
- First-quarter Moon
- Full Moon
- Last-quarter Moon

Look at the calendar above, using your data from the previous page, write the dates of the phases.

1. New Moon ______________________
2. New Crescent ______________________
3. First Quarter ______________________
4. New Gibbous ______________________
5. Full Moon ______________________
6. Old Gibbous ______________________
7. Last Quarter ______________________
8. Old Crescent ______________________

Observations:

1. What is a synodic month?

____

____

2. Does the Moon's shadow change relative to the Sun?

____

3. Which side is lit when the Moon is waxing, getting larger?

____

4. What is it called as a Moon appears to get larger?

5. During which lunar phase do we see only 50% of the Moon lit?

6. Why is there a difference between a synodic month and a sidereal month?

7. On which date will the next first-quarter moon phase occur?

8. What phase is likely to occur on July 19?

9. Why do we always see the same side of the Moon?

10. Pretend your lab partner was absent. Describe how lunar phases change.

38 Tides and Time

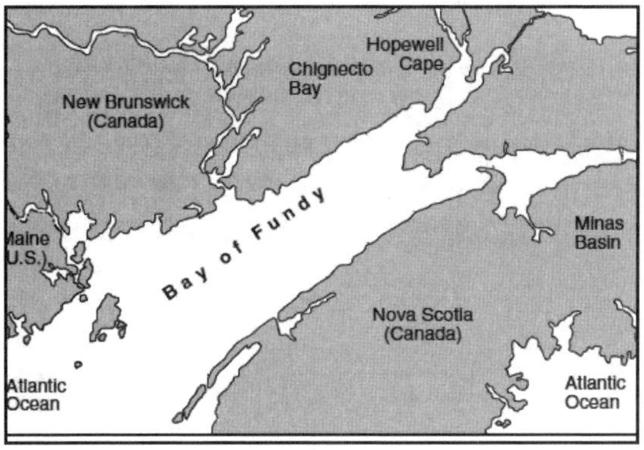

There is an old saying, "Tides and time wait for no man." It means that tides are constant and predictable and most importantly, unstoppable. Tides are the daily changes in the surface level of the Earth's oceans. Each quarter rotation of the Earth causes a major change in tides.

Different areas of the planet experience tides differently. Some areas have almost no tidal changes and in other areas the tides are dramatic. The Bay of Fundy in Canada has the highest tidal range on Earth.

Tides are caused by the pull of the Moon's gravity. Each day, there are two high tides and two low tides. Because the Moon is closer to the earth than the sun is, the Moon has a stronger gravitational pull on the Earth. During the full Moon and new moon phases, spring tides occur. Spring tides are simply the highest of high tides and the lowest of low tides. During the first-quarter and third-quarter phases of the Moon, neap tides occur. Neap tides don't create a great tidal range, thus we experience the smallest high tides and low tides.

Objective:

The objective of this lab is to plot a tide at the Battery in New York Harbor to determine if a pattern exists in tides.

Essential Question:

____

____

Hypothesis:

If ____

Then ____

Because ____

Variables:

Independent: ____

Dependent: ____

Materials: • Tide chart • Tidal Ranges – New York Harbor chart

Procedure:

1. Find the tidal predictions in the Tidal Ranges – New York Harbor chart below.
2. Plot the dates on the horizontal axis of the Tidal Changes graph located on next page.
3. Plot the water level on the vertical axis.
4. Review Lunar Orbit and Phases investigation (pages 185-188).

The following is a tidal chart, recorded hourly at the area of New York Harbor known as the Battery, the southern tip of Manhattan Island. It is named the Battery for the artillery battery that protected the early settlement of New York.

Tidal Ranges – New York Harbor

Date	Hours	Water Level (meters)
April 10	**00:00:00 am**	**1.4**
April 10	01:00:00 am	1.2
April 10	02:00:00 am	1.0
April 10	03:00:00 am	0.8
April 10	04:00:00 am	0.6
April 10	05:00:00 am	0.5
April 10	06:00:00 am	1.0
April 10	07:00:00 am	1.3
April 10	08:00:00 am	1.7
April 10	09:00:00 am	1.9
April 10	10:00:00 am	2.0
April 10	11:00:00 am	1.7
April 10	12:00:00 pm	1.4
April 10	01:00:00 pm	1.0
April 10	02:00:00 pm	0.7
April 10	03:00:00 pm	0.4
April 10	04:00:00 pm	0.5
April 10	05:00:00 pm	0.6
April 10	06:00:00 pm	0.9
April 10	07:00:00 pm	1.4
April 10	08:00:00 pm	1.8
April 10	09:00:00 pm	2.0
April 10	10:00:00 pm	2.2
April 10	11:00:00 pm	1.9
April 11	**00:00:00 am**	**1.6**
April 11	01:00:00 am	1.2
April 11	02:00:00 am	0.8
April 11	03:00:00 am	0.7
April 11	04:00:00 am	0.5
April 11	05:00:00 am	0.5
April 11	06:00:00 am	0.7
April 11	07:00:00 am	1.1
April 11	08:00:00 am	1.5
April 11	09:00:00 am	1.7
April 11	10:00:00 am	1.8
April 11	11:00:00 am	1.8
April 11	12:00:00 pm	1.6
April 11	01:00:00 pm	1.2
April 11	02:00:00 pm	0.9
April 11	03:00:00 pm	0.6
April 11	04:00:00 pm	0.4
April 11	05:00:00 pm	0.5
April 11	06:00:00 pm	0.7
April 11	07:00:00 pm	1.2
April 11	08:00:00 pm	1.6
April 11	09:00:00 pm	1.9
April 11	10:00:00 pm	2.1
April 11	11:00:00 pm	2.0
April 12	**00:00:00 am**	**1.8**
April 12	01:00:00 am	1.5
April 12	02:00:00 am	1.1
April 12	03:00:00 am	0.8
April 12	04:00:00 am	0.5
April 12	05:00:00 am	0.4
April 12	06:00:00 am	0.5
April 12	07:00:00 am	0.8
April 12	08:00:00 am	1.2
April 12	09:00:00 am	1.5
April 12	10:00:00 am	1.7

Tidal Changes

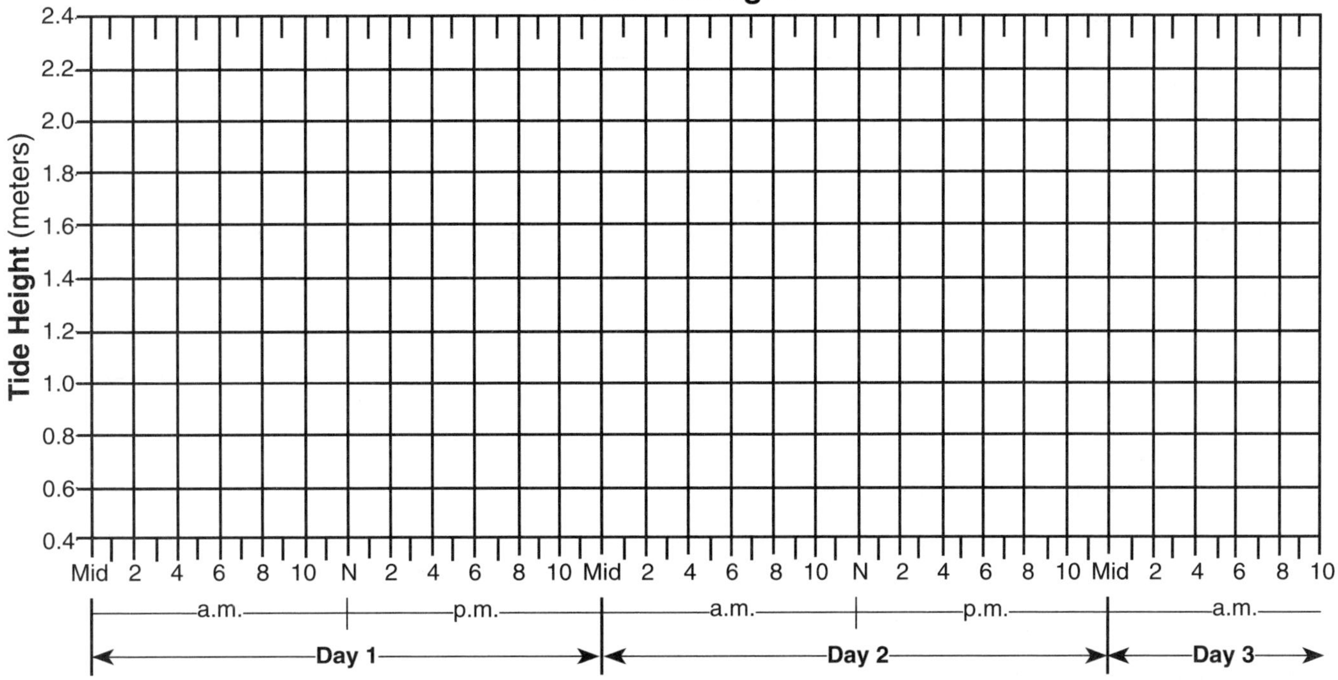

Key
Mid – Midnight N – Noon

Observations:

1. Is your graph above indicative of a random, direct, inverse or cyclic pattern?

____

____

2. Approximately, how much time elapses between high tides?

____

____

3. Is one high tide higher than another in a twenty-four hour period? Why?

____

____

4. If a complete cycle is from high tide to high tide, how many complete cycles are on your graph?

____

____

5. What is the time of the first low tide on your graph?

6. What is the average tidal range on the Tidal Ranges – New York Harbor chart?

7. Is the tidal range increasing or decreasing?

8. What causes tides?

9. Why are there two tidal cycles in a twenty-four hour period?

10. Assuming the Tidal Ranges – New York Harbor chart reflects the maximum range, what can you infer about the lunar cycle in your graph?

NAME ______________________

39 Newton's Laws

Sir Isaac Newton was an English physicist, mathematician, and philosopher. Newton is thought to be one of the most important scientists of all time. Newton formulated laws of gravity and motion—laws that explain how objects move on Earth as well as through the sky. His mathematical insights led him to invent the area of mathematics called calculus.

Newton formulated what are now called Newton's Three Laws of Motions. These laws are based on careful observations and by extensive mathematical proofs. Popular lore describes Newton theorizing the laws of gravitation and motion after being hit on the head from a falling apple. As a result of his work, the unit to measure force of work is called a Newton.

Newton's First Law

I. Objects in motion tend to stay in motion and objects at rest tend to stay at rest unless acted upon by an outside force.

The Second Law

II. An object with a certain velocity maintains that velocity *unless* a force acts on it to cause acceleration or deceleration.

The Third Law

III. For every action there is an equal and opposite reaction.

Objective:

The objective of this lab is to make observations and to prove Newton's Law of Motion.

Essential Question:

____

____

Materials:

- Model rockets
- A and D Engines
- Launch pad
- Safety tape or safety cones

The Third Law

III. For every action there is an equal and opposite reaction.

At the completion of this activity, the class will go outside for a demonstration of Newton's Third Law.

In this task, the Newtons of force of two rocket engines will be compared. The rocket will react to an action and equal and opposite reaction.

Procedure:

Create a safe zone for the rocket launch. Record the Newtons on the rocket engine.

Do not cross the designated safe zone for the rocket launch.

Theorize: Which will exert more force, the higher or lower number of Newtons listed on the engine?

____

____

____

____

____

____

____

____

Engine:	Newtons:	Observations:
A		
D		

Observations:

1. Which trial demonstrated a greater release of energy, the first or second trial?

___

___

2. What were the Newtons on the respective engines?

___

___

3. How did that affect the force of energy applied to the rocket?

___

___

4. What were some unexpected events during this lab activity?

___

___

5. If objects in motion tend to stay in motion, what force or forces act to slow down objects in motion?

___

___

6. If you were in a gym class and your teacher asked you to push down on the floor, why are they called "push-ups"?

___

___

7. Thinking of question number 6, which is exerting more Newtons, you or the floor?

___

___

8. Thinking about question number 5, explain why it is unsafe to drive quickly on ice.

___

___

9. What keeps the planets orbiting the Sun?

10. What keeps the planets in their orbit?

NAME ____________________

40 Kepler's Laws

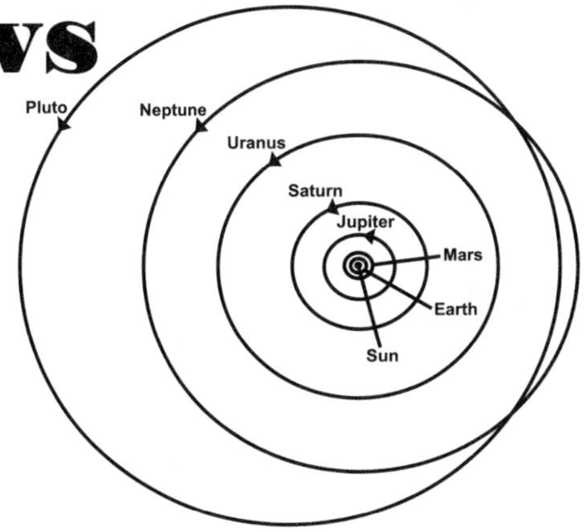

Johannes Kepler was a German mathematician and astronomer. He had studied astronomy long before he met Tycho Brahe. Johannes Kepler worked for Tycho Brahe as an assistant in Prague. Tycho Brahe is considered the most accomplished pre-telescope astronomer. Brahe had started tables of planetary orbits. Kepler continued the work on the tables and was unable to have the tables be reconciled within a circular orbit. He realized that planet's orbits are elliptical. This is Kepler's first law. He also agreed with the Copernican world-view. It is a heliocentric model of planetary motion.

An ellipse has an oval shape. The earth travels around the sun in an elliptical orbit. Orbital velocity is the speed at which a planet travels in its orbit. The Earth's orbital velocity is greatest at perihelion and least at aphelion. Gravity pulls all objects toward the center of the earth. The downward pull of gravity causes a thrown ball to follow a curved path. There is a gravitational attraction between every object in the universe. The Sun's gravitational attraction keeps the planets moving in elliptical orbits and determines their orbital velocities.

Kepler's First Law

I. Planets move along ellipses, or ovals, with the Sun at one focus.

Kepler's Second Law

II. The line from the Sun to the planet covers equal areas in equal times.

Kepler's Third Law

III. The square of the planet's orbital period is proportional to the third power ("cube") of the average distance from the Sun.

Objective:

The objective of this lab is to make observations and to prove Kepler's First and Third Law of Motion.

Essential Question: ____________________

____

Hypothesis:

If ____________________

Then ____________________

Because ____________________

Variables:

Independent:____________________________

Dependent:____________________________

Materials:

- Cotton string (about 25 cm)
- Foam board (about 22 cm x 27 cm)
- A small container to hold push pins
- Four-Function calculator
- Earth Science Reference Tables in the back of your lab workbook
- 2 push pins
- Felt tip marker
- Tape
- Metric ruler

Procedure:

1. Tear out the Foci Page (page 201).
2. Tape the page to the foam board.
3. Insert the push pins into the dots on the page in figure 1.
4. Place the string around the push pins.
5. Using the felt tip marker, draw a shape by having the marker pull the string to the limits the string will move between the push pins.
6. Repeat the above procedure for figure 2. Use the same string.

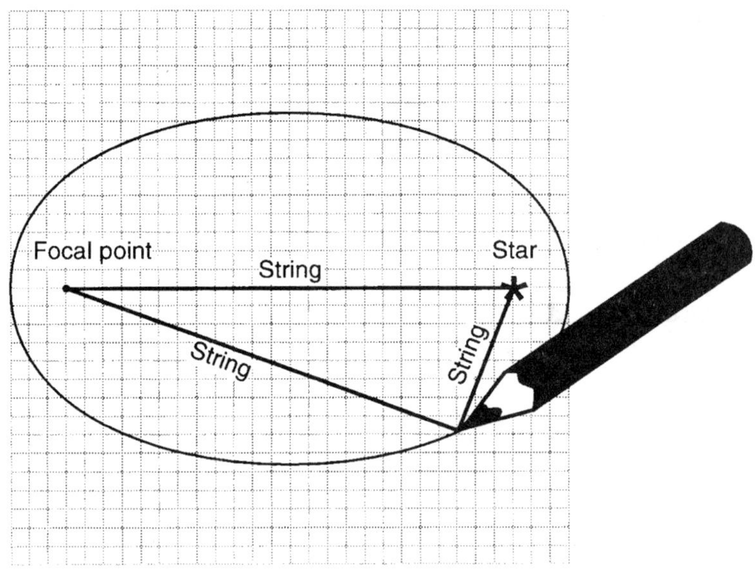

(Drawn to scale)

Observations:

1. What is the length of the major axis in figure 1?

____

2. What is the distance between the two foci in figure 1?

____

3. Determine the eccentricity of the ellipse.

____

4. What is the length of the major axis in figure 2?

____

5. What is the distance between the two foci in figure 2?

____

6. Determine the eccentricity of the ellipse.

____

7. Compare Figure 1 and Figure 2. How are the ellipses different?

____

____

8. Compare Figure 1 and Figure 2. How are the ellipses similar?

____

____

9. Which ellipse is closer to being a circle?

____

____

10. Without being able to show someone a picture of an ellipse, how could you determine its eccentricity?

____

____

____

____

____

____

____

____

Foci Page

– Tear out this page –

– Figure 1 –

Distance between foci: _____ cm. Length of major axis _____ cm

– Figure 2 –

Distance between foci: _____ cm. Length of major axis _____ cm

Reference Tables for Physical Setting/EARTH SCIENCE 2011 Edition

Radioactive Decay Data

RADIOACTIVE ISOTOPE	DISINTEGRATION	HALF-LIFE (years)
Carbon-14	$^{14}C \rightarrow {}^{14}N$	5.7×10^3
Potassium-40	$^{40}K \rightarrow {}^{40}Ar$, $^{40}K \rightarrow {}^{40}Ca$	1.3×10^9
Uranium-238	$^{238}U \rightarrow {}^{206}Pb$	4.5×10^9
Rubidium-87	$^{87}Rb \rightarrow {}^{87}Sr$	4.9×10^{10}

Specific Heats of Common Materials

MATERIAL	SPECIFIC HEAT (Joules/gram • °C)
Liquid water	4.18
Solid water (ice)	2.11
Water vapor	2.00
Dry air	1.01
Basalt	0.84
Granite	0.79
Iron	0.45
Copper	0.38
Lead	0.13

Equations

$$\text{Eccentricity} = \frac{\text{distance between foci}}{\text{length of major axis}}$$

$$\text{Gradient} = \frac{\text{change in field value}}{\text{distance}}$$

$$\text{Rate of change} = \frac{\text{change in value}}{\text{time}}$$

$$\text{Density} = \frac{\text{mass}}{\text{volume}}$$

Properties of Water

Heat energy gained during melting . . . . 334 J/g
Heat energy released during freezing . . . . 334 J/g
Heat energy gained during vaporization . . . . . 2260 J/g
Heat energy released during condensation . . . 2260 J/g

Density at 3.98°C . . . . 1.0 g/mL

Average Chemical Composition of Earth's Crust, Hydrosphere, and Troposphere

ELEMENT (symbol)	CRUST		HYDROSPHERE	TROPOSPHERE
	Percent by mass	Percent by volume	Percent by volume	Percent by volume
Oxygen (O)	46.10	94.04	33.0	21.0
Silicon (Si)	28.20	0.88		
Aluminum (Al)	8.23	0.48		
Iron (Fe)	5.63	0.49		
Calcium (Ca)	4.15	1.18		
Sodium (Na)	2.36	1.11		
Magnesium (Mg)	2.33	0.33		
Potassium (K)	2.09	1.42		
Nitrogen (N)				78.0
Hydrogen (H)			66.0	
Other	0.91	0.07	1.0	1.0

Generalized Landscape Regions of New York State

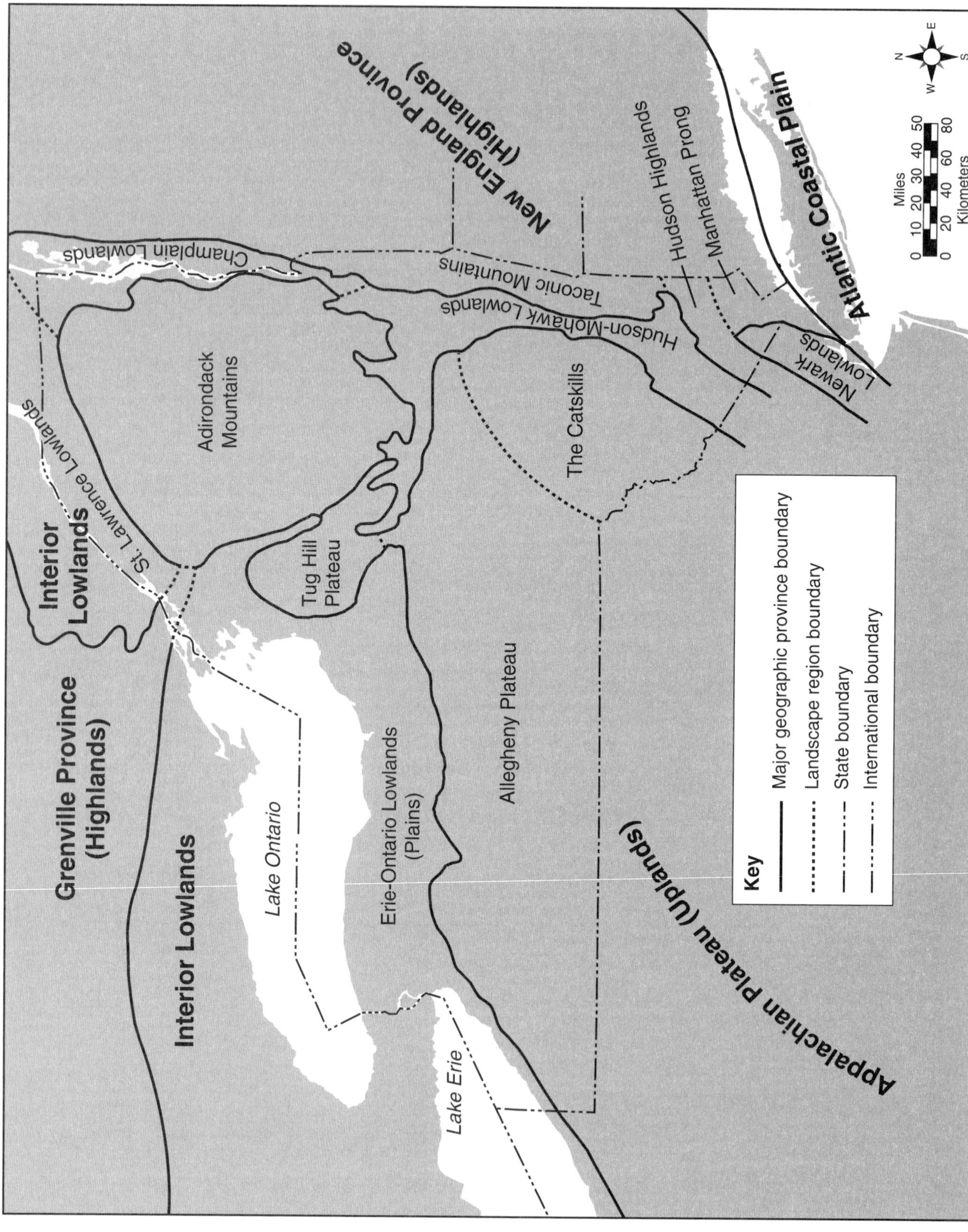

Generalized Bedrock Geology of New York State

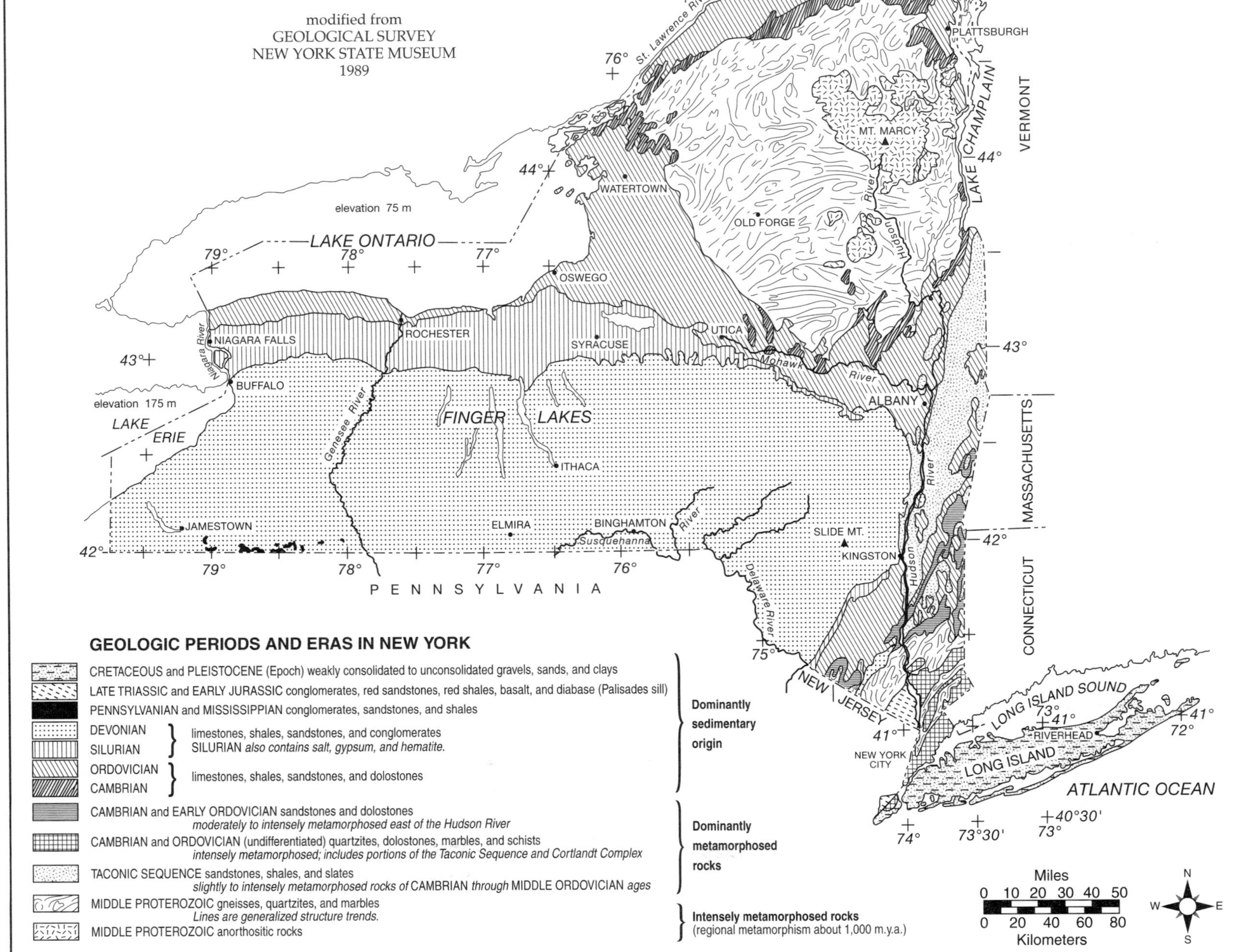

Surface Ocean Currents

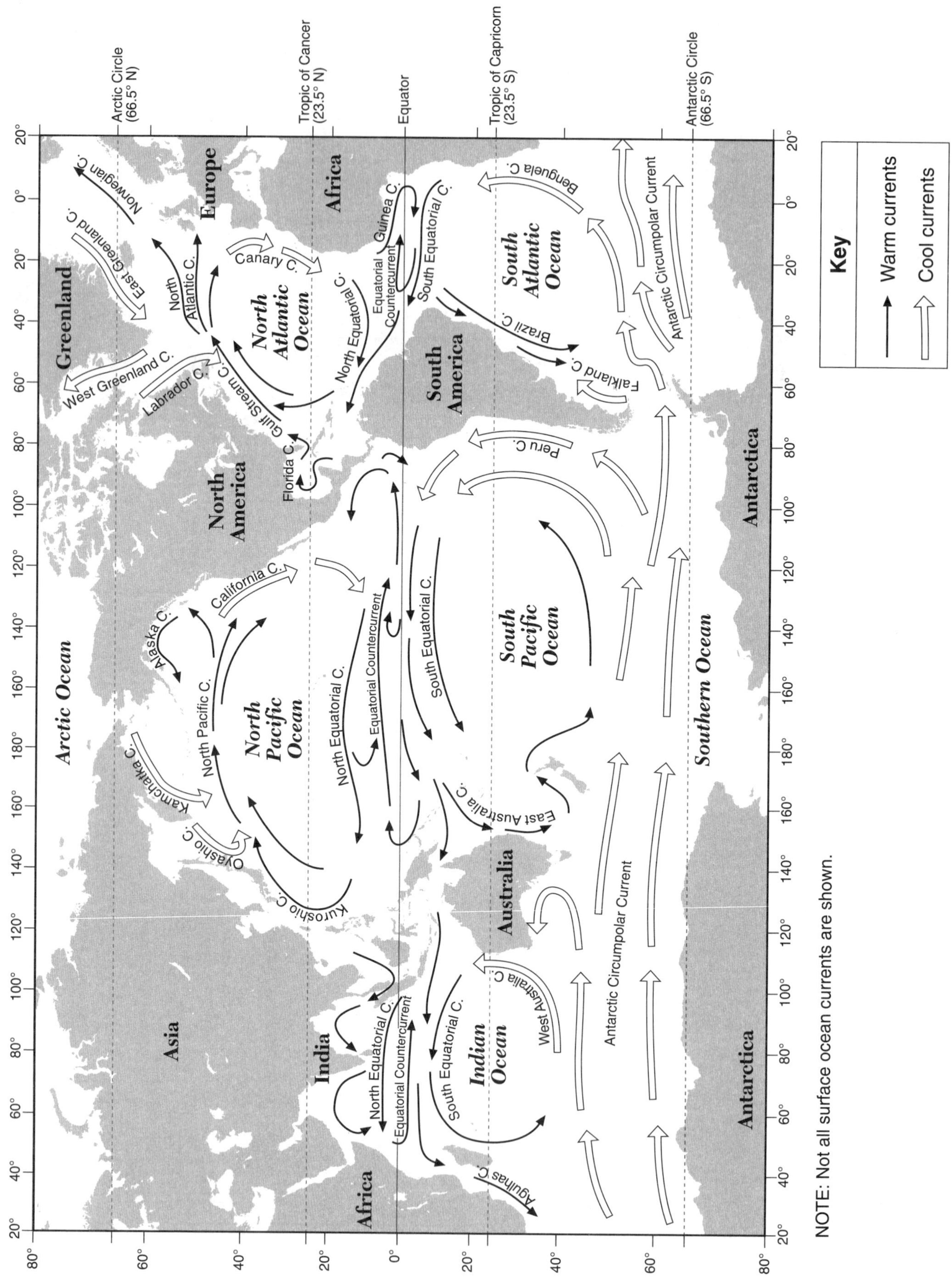

NOTE: Not all surface ocean currents are shown.

Tectonic Plates

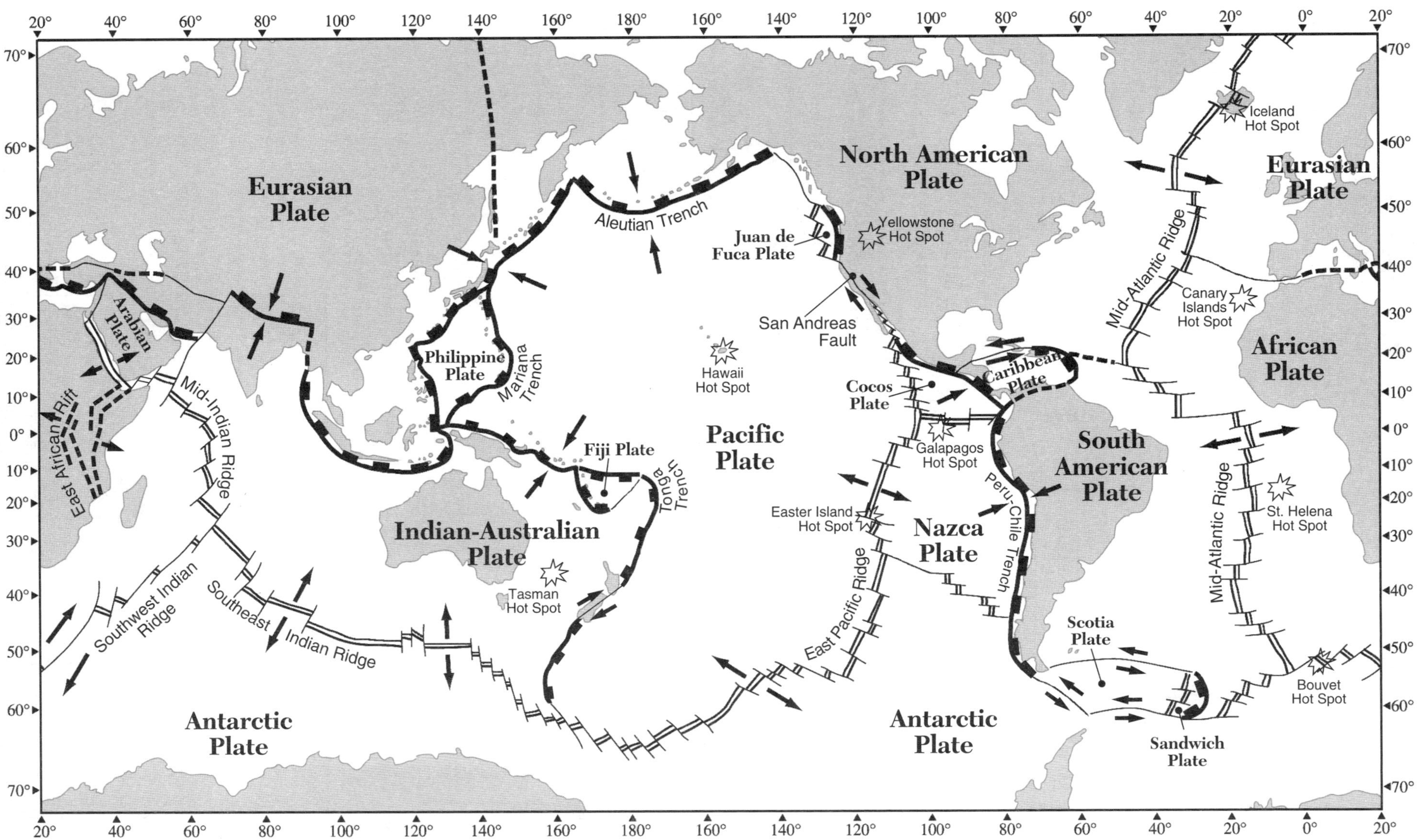

Key

- Relative motion at plate boundary
- Transform plate boundary (transform fault)
- Divergent plate boundary (usually broken by transform faults along mid-ocean ridges)
- Convergent plate boundary (subduction zone) — overriding plate / subducting plate
- Complex or uncertain plate boundary
- Mantle hot spot

NOTE: Not all mantle hot spots, plates, and boundaries are shown.

Rock Cycle in Earth's Crust

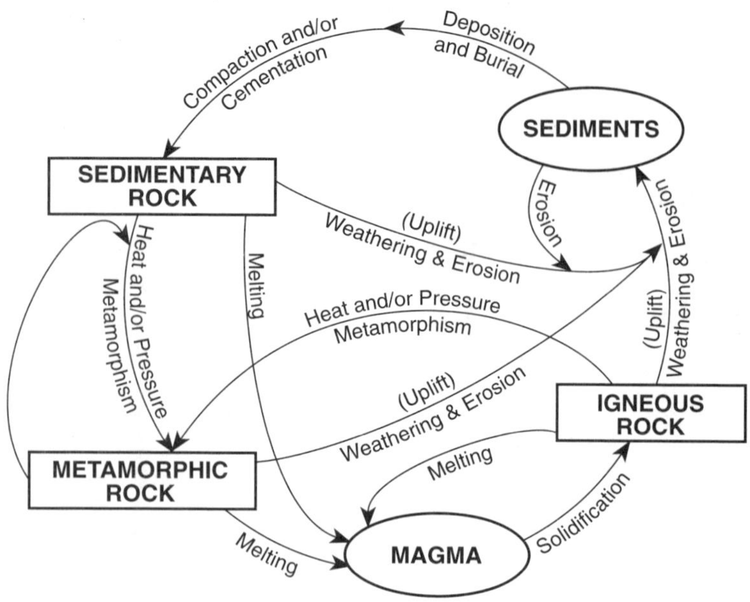

Relationship of Transported Particle Size to Water Velocity

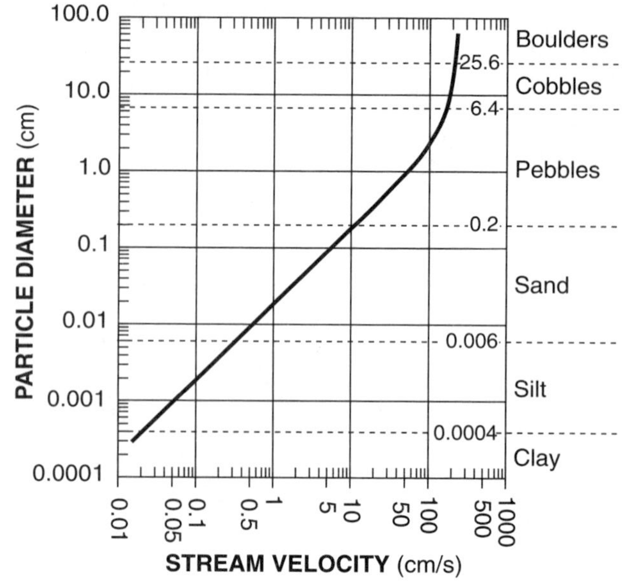

This generalized graph shows the water velocity needed to maintain, but not start, movement. Variations occur due to differences in particle density and shape.

Scheme for Igneous Rock Identification

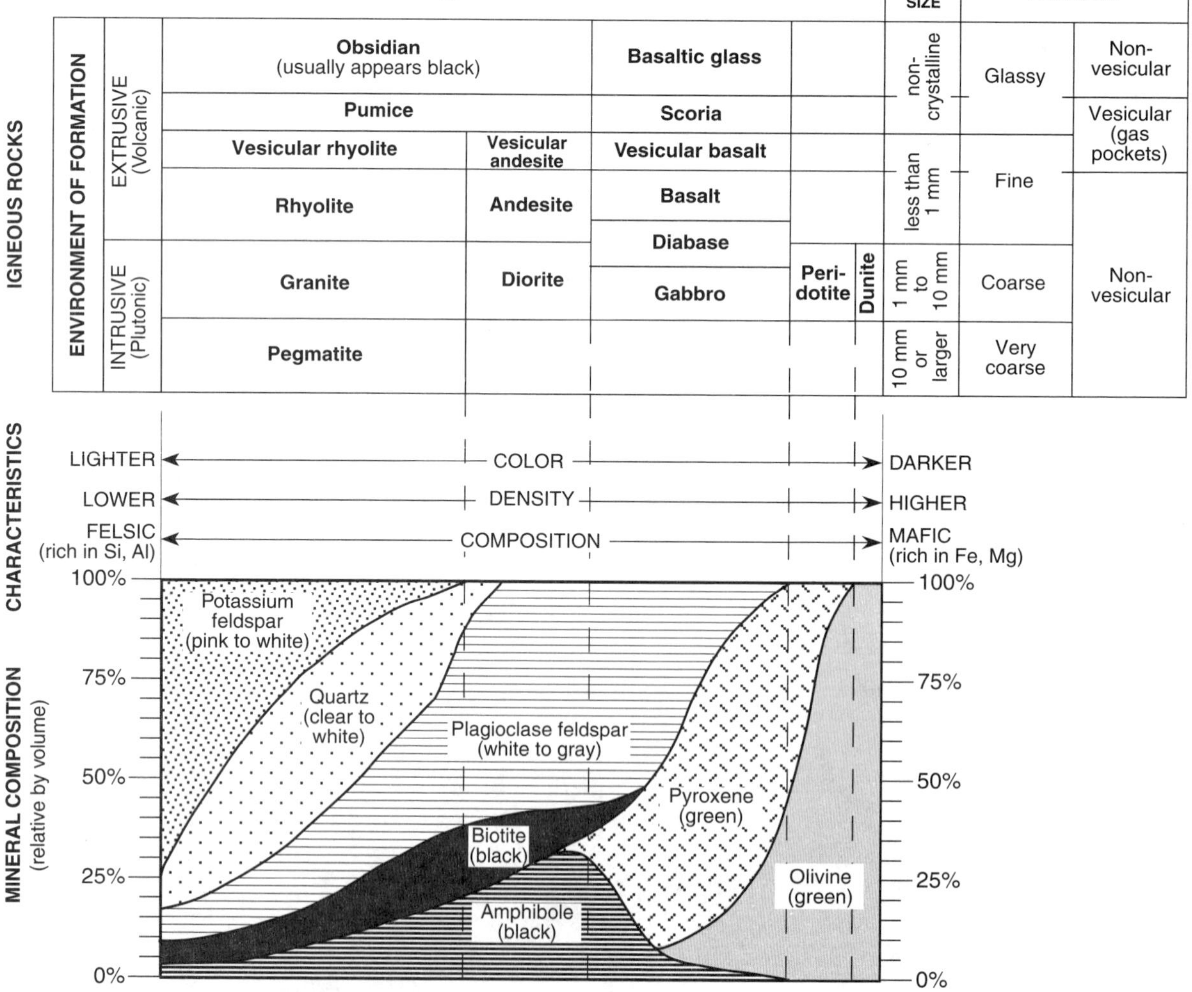

Scheme for Sedimentary Rock Identification

INORGANIC LAND-DERIVED SEDIMENTARY ROCKS					
TEXTURE	**GRAIN SIZE**	**COMPOSITION**	**COMMENTS**	**ROCK NAME**	**MAP SYMBOL**
Clastic (fragmental)	Pebbles, cobbles, and/or boulders embedded in sand, silt, and/or clay	Mostly quartz, feldspar, and clay minerals; may contain fragments of other rocks and minerals	Rounded fragments	**Conglomerate**	
			Angular fragments	**Breccia**	
	Sand (0.006 to 0.2 cm)		Fine to coarse	**Sandstone**	
	Silt (0.0004 to 0.006 cm)		Very fine grain	**Siltstone**	
	Clay (less than 0.0004 cm)		Compact; may split easily	**Shale**	

CHEMICALLY AND/OR ORGANICALLY FORMED SEDIMENTARY ROCKS					
TEXTURE	**GRAIN SIZE**	**COMPOSITION**	**COMMENTS**	**ROCK NAME**	**MAP SYMBOL**
Crystalline	Fine to coarse crystals	Halite	Crystals from chemical precipitates and evaporites	**Rock salt**	
		Gypsum		**Rock gypsum**	
		Dolomite		**Dolostone**	
Crystalline or bioclastic	Microscopic to very coarse	Calcite	Precipitates of biologic origin or cemented shell fragments	**Limestone**	
Bioclastic		Carbon	Compacted plant remains	**Bituminous coal**	

Scheme for Metamorphic Rock Identification

TEXTURE		**GRAIN SIZE**	**COMPOSITION**	**TYPE OF METAMORPHISM**	**COMMENTS**	**ROCK NAME**	**MAP SYMBOL**
FOLIATED	MINERAL ALIGNMENT	Fine	MICA	Regional (Heat and pressure increases) ↓	Low-grade metamorphism of shale	**Slate**	
		Fine to medium	MICA, QUARTZ, FELDSPAR, AMPHIBOLE, GARNET		Foliation surfaces shiny from microscopic mica crystals	**Phyllite**	
			MICA, QUARTZ, FELDSPAR, AMPHIBOLE, GARNET, PYROXENE		Platy mica crystals visible from metamorphism of clay or feldspars	**Schist**	
	BAND-ING	Medium to coarse	MICA, QUARTZ, FELDSPAR, AMPHIBOLE, GARNET, PYROXENE		High-grade metamorphism; mineral types segregated into bands	**Gneiss**	
NONFOLIATED		Fine	Carbon	Regional	Metamorphism of bituminous coal	**Anthracite coal**	
		Fine	Various minerals	Contact (heat)	Various rocks changed by heat from nearby magma/lava	**Hornfels**	
		Fine to coarse	Quartz	Regional or contact	Metamorphism of quartz sandstone	**Quartzite**	
			Calcite and/or dolomite		Metamorphism of limestone or dolostone	**Marble**	
		Coarse	Various minerals		Pebbles may be distorted or stretched	**Metaconglomerate**	

GEOLOGIC HISTORY

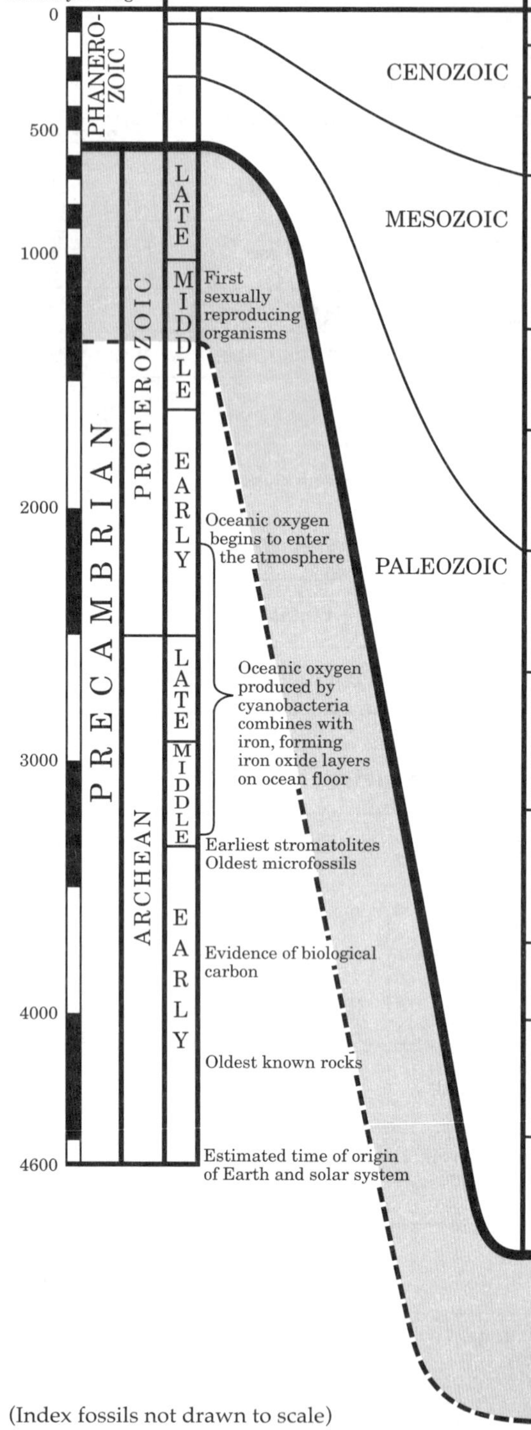

Era	Period		Epoch	Million years ago	Life on Earth
CENOZOIC	*QUATERNARY*		HOLOCENE	0 – 0.01	
			PLEISTOCENE	1.8	Humans, mastodonts, mammoths
	NEOGENE		PLIOCENE	5.3	Large carnivorous mammals
			MIOCENE	23.0	Abundant grazing mammals
	PALEOGENE		OLIGOCENE	33.9	Earliest grasses
			EOCENE	55.8	Many modern groups of mammals
			PALEOCENE	65.5	Mass extinction of dinosaurs, ammonoids, and many land plants
MESOZOIC	*CRETACEOUS*		LATE		
			EARLY	146	Earliest flowering plants Diverse bony fishes
	JURASSIC		LATE		Earliest birds
			MIDDLE		Abundant dinosaurs and ammonoids
			EARLY	200	
	TRIASSIC		LATE		Earliest mammals Earliest dinosaurs
			MIDDLE		
			EARLY	251	Mass extinction of many land and marine organisms (including trilobites)
PALEOZOIC	*PERMIAN*		LATE		
			MIDDLE		Mammal-like reptiles
			EARLY	299	Abundant reptiles
	CARBONIFEROUS	*PENNSYLVANIAN*	LATE		
			EARLY	318	Extensive coal-forming forests
		MISSISSIPPIAN	LATE		Abundant amphibians
			MIDDLE		Large and numerous scale trees and seed ferns (vascular plants); earliest reptiles
			EARLY	359	
	DEVONIAN		LATE		Earliest amphibians and plant seeds Extinction of many marine organisms
			MIDDLE		Earth's first forests Earliest ammonoids and sharks
			EARLY	416	Abundant fish
	SILURIAN		LATE		Earliest insects Earliest land plants and animals
			EARLY	444	Abundant eurypterids
	ORDOVICIAN		LATE		
			MIDDLE		Invertebrates dominant Earth's first coral reefs
			EARLY	488	
	CAMBRIAN		LATE		
			MIDDLE		Burgess shale fauna (diverse soft-bodied organisms) Earliest fishes
			EARLY	542	Extinction of many primitive marine organisms Earliest trilobites Great diversity of life-forms with shelly parts
				580	Ediacaran fauna (first multicellular, soft-bodied marine organisms)
				1300	Abundant stromatolites

NY Rock Record: Sediment; Bedrock

(Index fossils not drawn to scale)

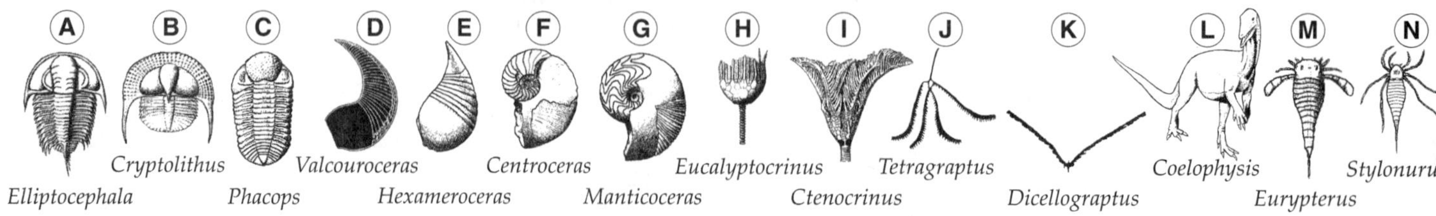

OF NEW YORK STATE

Time Distribution of Fossils (including important fossils of New York)

The center of each lettered circle indicates the approximate time of existence of a specific index fossil (e.g. Fossil (A) lived at the end of the Early Cambrian).

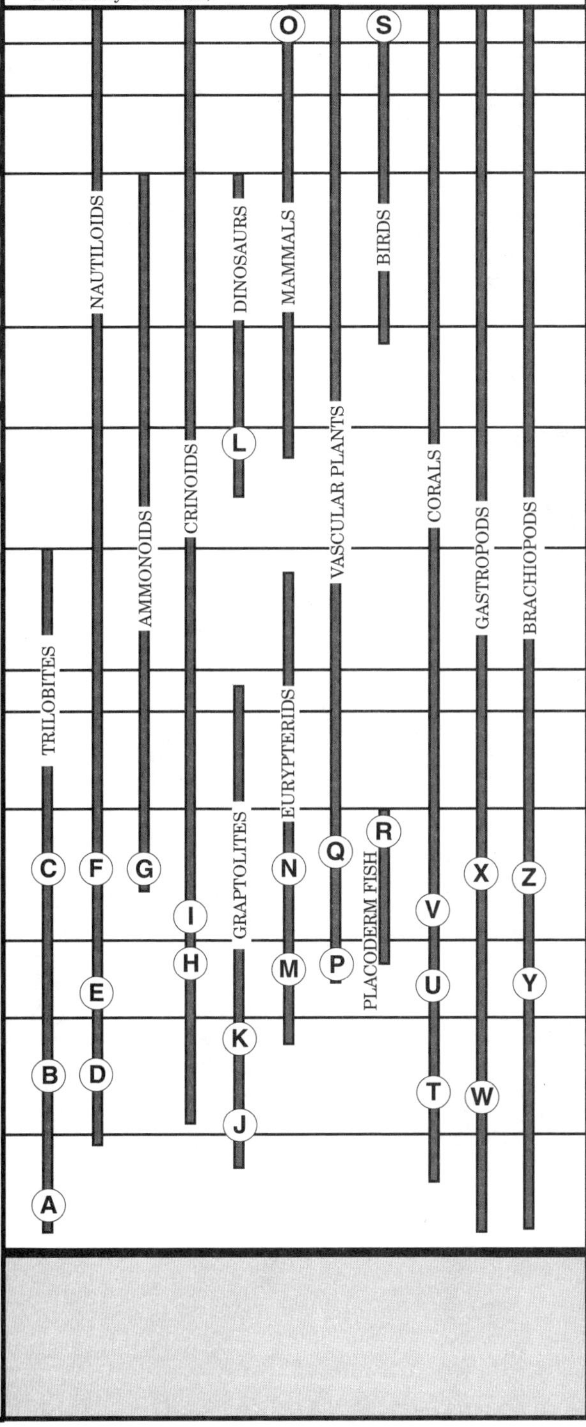

Important Geologic Events in New York

- Advance and retreat of last continental ice
- Sands and clays underlying Long Island and Staten Island deposited on margin of Atlantic Ocean
- Dome-like uplift of Adirondack region begins
- Initial opening of Atlantic Ocean
- North America and Africa separate
- Intrusion of Palisades sill
- Pangaea begins to break up
- **Alleghenian orogeny** caused by collision of North America and Africa along transform margin, forming Pangaea
- Catskill delta forms
- Erosion of Acadian Mountains
- **Acadian orogeny** caused by collision of North America and Avalon and closing of remaining part of Iapetus Ocean
- Salt and gypsum deposited in evaporite basins
- Erosion of Taconic Mountains; Queenston delta forms
- **Taconian orogeny** caused by closing of western part of Iapetus Ocean and collision between North America and volcanic island arc
- Widespread deposition over most of New York along edge of Iapetus Ocean
- Rifting and initial opening of Iapetus Ocean
- Erosion of Grenville Mountains
- **Grenville orogeny:** metamorphism of bedrock now exposed in the Adirondacks and Hudson Highlands

Inferred Positions of Earth's Landmasses

59 million years ago

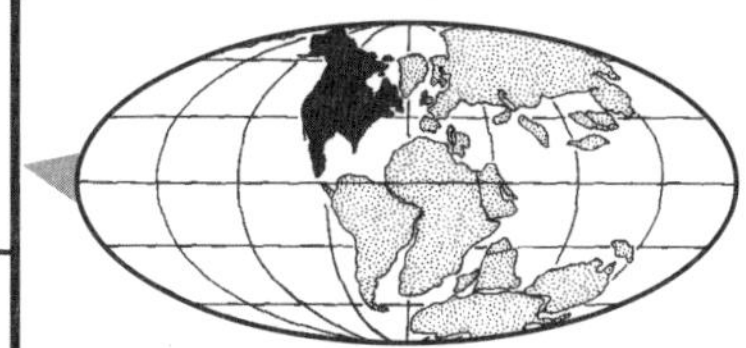

119 million years ago

232 million years ago

359 million years ago

458 million years ago

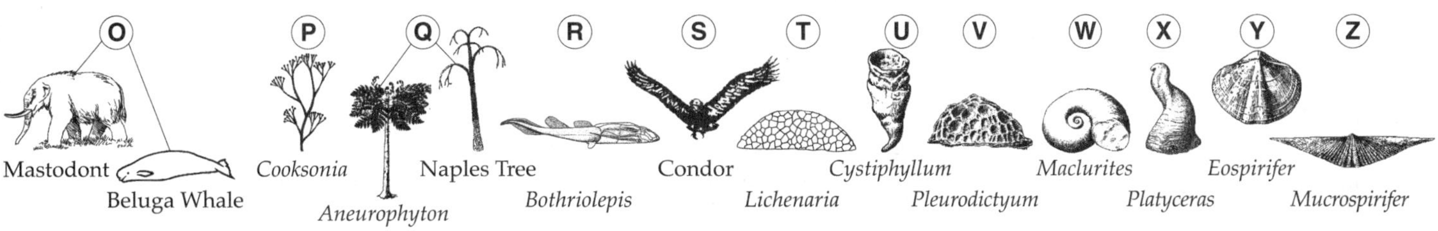

Inferred Properties of Earth's Interior

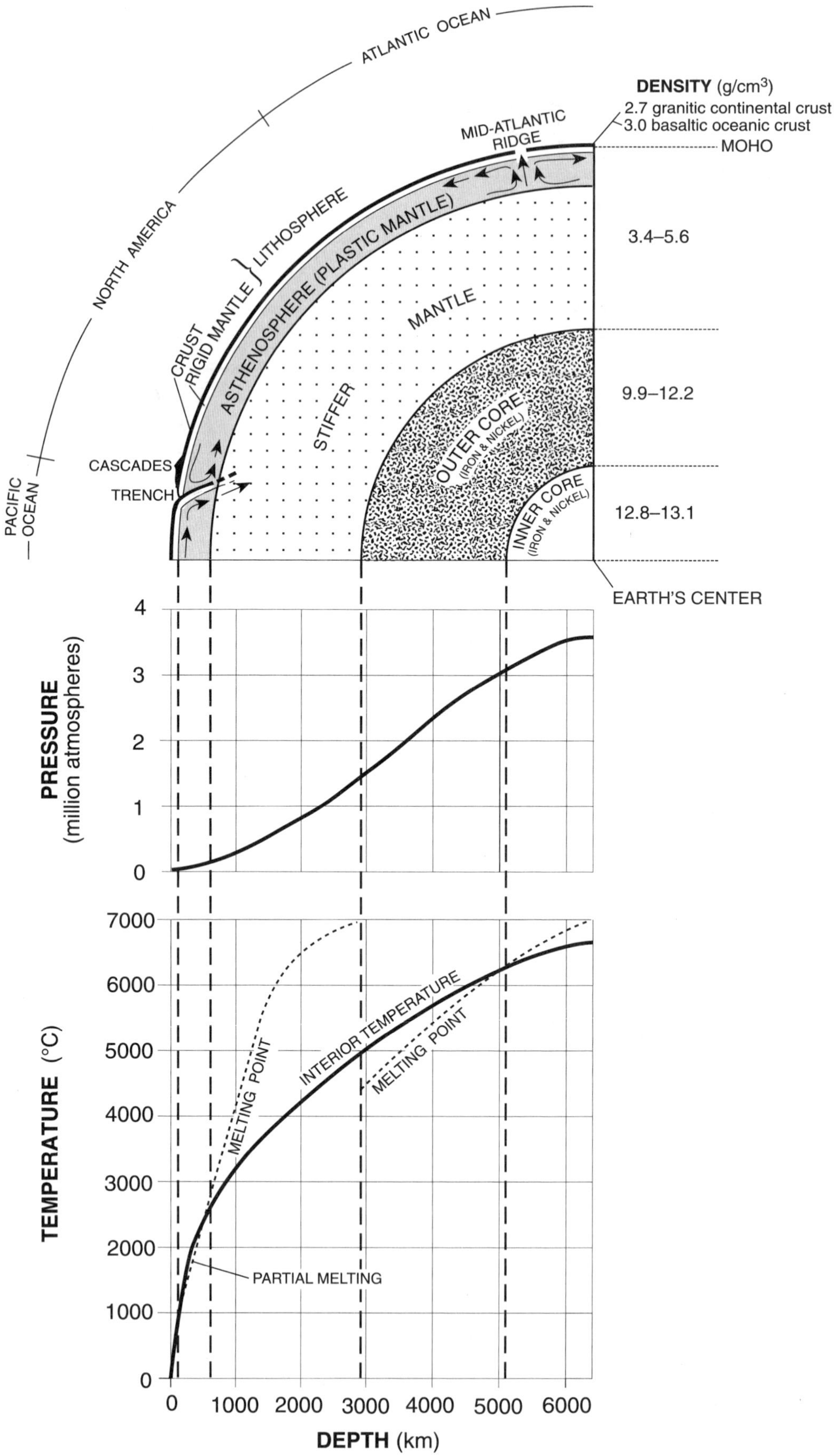

Earthquake P-Wave and S-Wave Travel Time

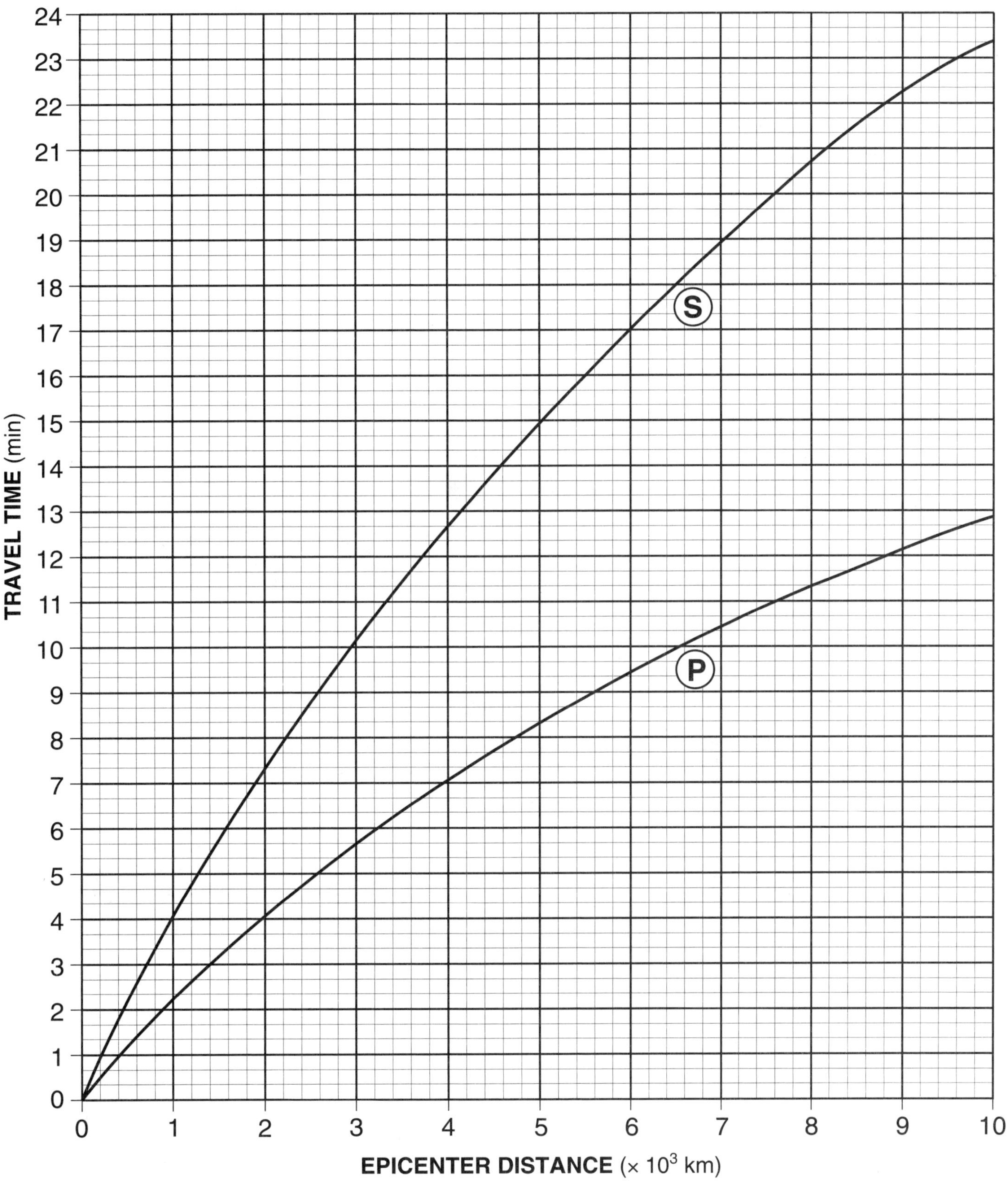

Dewpoint (°C)

Dry-Bulb Tempera-ture (°C)	Difference Between Wet-Bulb and Dry-Bulb Temperatures (C°)															
	0	1	2	3	4	5	6	7	8	9	10	11	12	13	14	15
−20	−20	−33														
−18	−18	−28														
−16	−16	−24														
−14	−14	−21	−36													
−12	−12	−18	−28													
−10	−10	−14	−22													
−8	−8	−12	−18	−29												
−6	−6	−10	−14	−22												
−4	−4	−7	−12	−17	−29											
−2	−2	−5	−8	−13	−20											
0	0	−3	−6	−9	−15	−24										
2	2	−1	−3	−6	−11	−17										
4	4	1	−1	−4	−7	−11	−19									
6	6	4	1	−1	−4	−7	−13	−21								
8	8	6	3	1	−2	−5	−9	−14								
10	10	8	6	4	1	−2	−5	−9	−14	−28						
12	12	10	8	6	4	1	−2	−5	−9	−16						
14	14	12	11	9	6	4	1	−2	−5	−10	−17					
16	16	14	13	11	9	7	4	1	−1	−6	−10	−17				
18	18	16	15	13	11	9	7	4	2	−2	−5	−10	−19			
20	20	19	17	15	14	12	10	7	4	2	−2	−5	−10	−19		
22	22	21	19	17	16	14	12	10	8	5	3	−1	−5	−10	−19	
24	24	23	21	20	18	16	14	12	10	8	6	2	−1	−5	−10	−18
26	26	25	23	22	20	18	17	15	13	11	9	6	3	0	−4	−9
28	28	27	25	24	22	21	19	17	16	14	11	9	7	4	1	−3
30	30	29	27	26	24	23	21	19	18	16	14	12	10	8	5	1

Relative Humidity (%)

Dry-Bulb Tempera-ture (°C)	Difference Between Wet-Bulb and Dry-Bulb Temperatures (C°)															
	0	1	2	3	4	5	6	7	8	9	10	11	12	13	14	15
−20	100	28														
−18	100	40														
−16	100	48														
−14	100	55	11													
−12	100	61	23													
−10	100	66	33													
−8	100	71	41	13												
−6	100	73	48	20												
−4	100	77	54	32	11											
−2	100	79	58	37	20	1										
0	100	81	63	45	28	11										
2	100	83	67	51	36	20	6									
4	100	85	70	56	42	27	14									
6	100	86	72	59	46	35	22	10								
8	100	87	74	62	51	39	28	17	6							
10	100	88	76	65	54	43	33	24	13	4						
12	100	88	78	67	57	48	38	28	19	10	2					
14	100	89	79	69	60	50	41	33	25	16	8	1				
16	100	90	80	71	62	54	45	37	29	21	14	7	1			
18	100	91	81	72	64	56	48	40	33	26	19	12	6			
20	100	91	82	74	66	58	51	44	36	30	23	17	11	5		
22	100	92	83	75	68	60	53	46	40	33	27	21	15	10	4	
24	100	92	84	76	69	62	55	49	42	36	30	25	20	14	9	4
26	100	92	85	77	70	64	57	51	45	39	34	28	23	18	13	9
28	100	93	86	78	71	65	59	53	47	42	36	31	26	21	17	12
30	100	93	86	79	72	66	61	55	49	44	39	34	29	25	20	16

Temperature

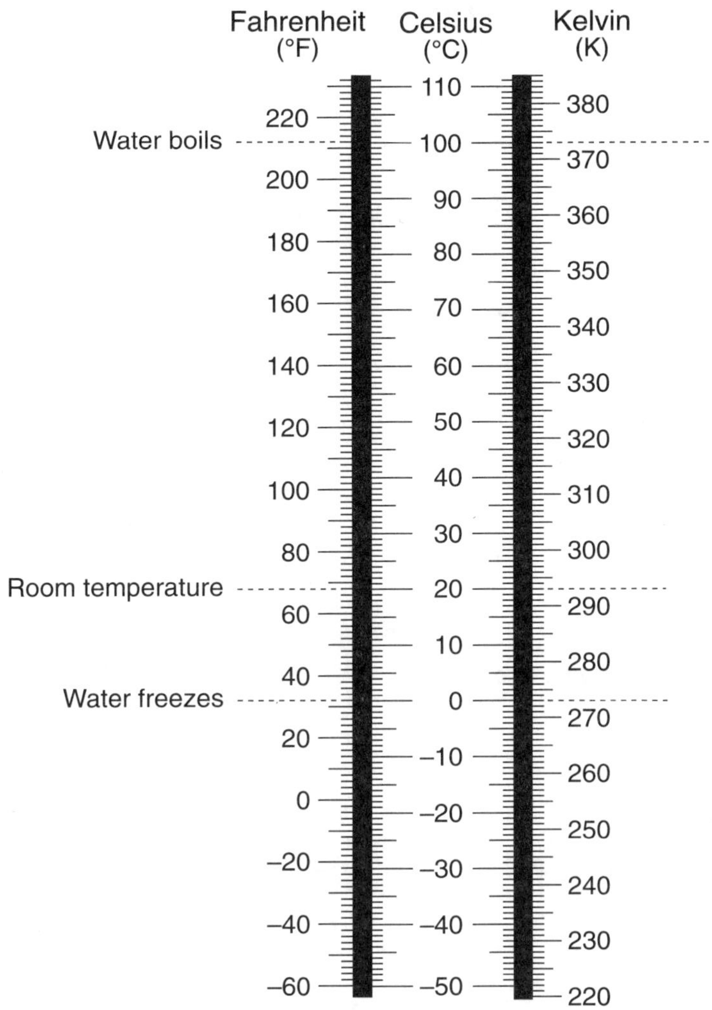

Pressure

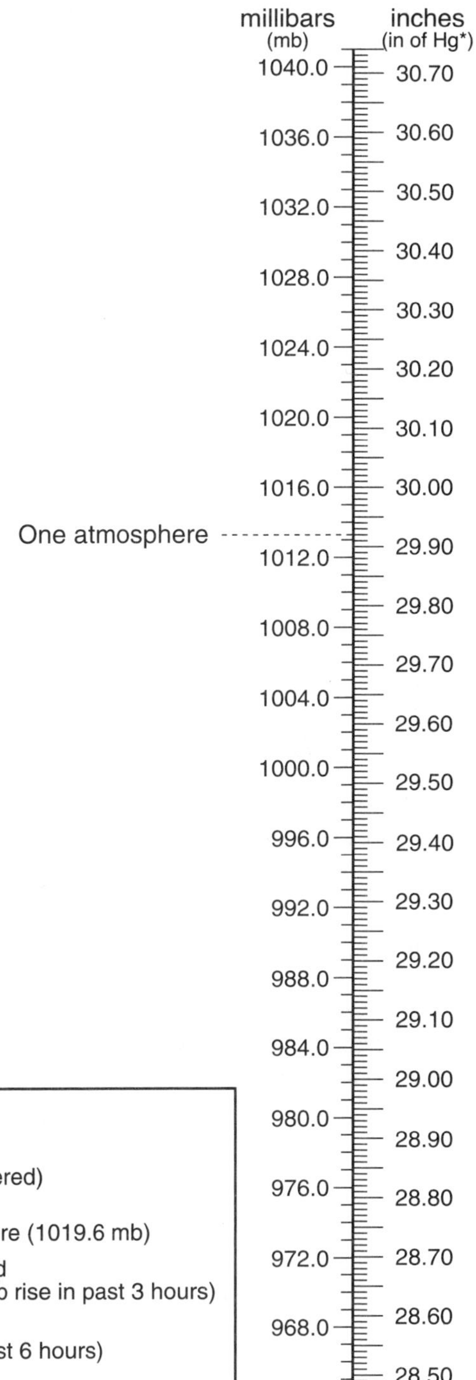

*Hg = mercury

Key to Weather Map Symbols

Station Model

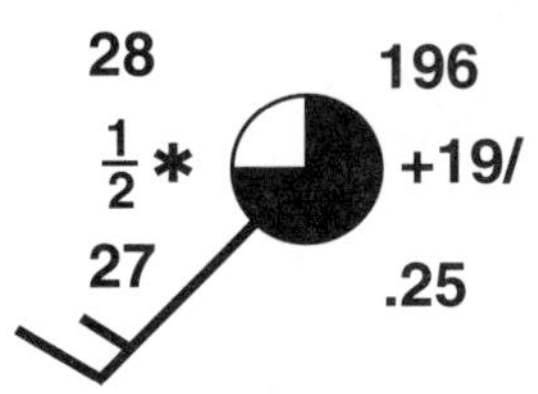

Station Model Explanation

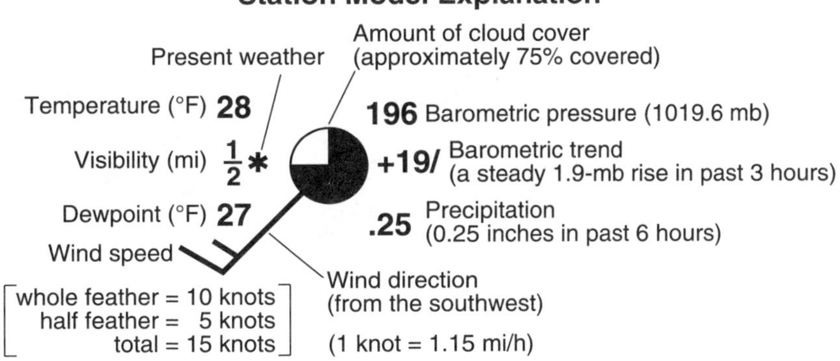

Present Weather

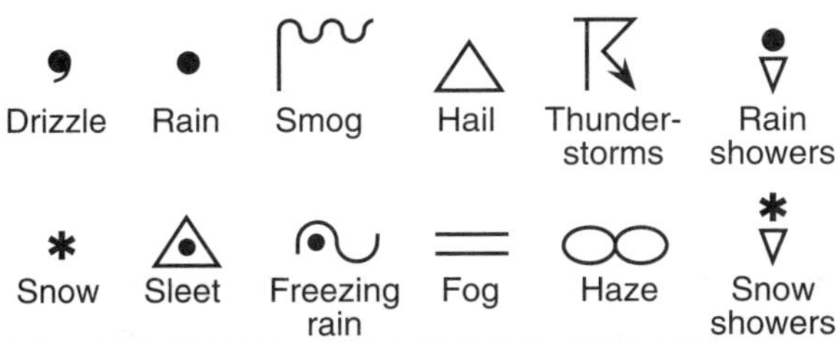

Air Masses

cA continental arctic
cP continental polar
cT continental tropical
mT maritime tropical
mP maritime polar

Fronts

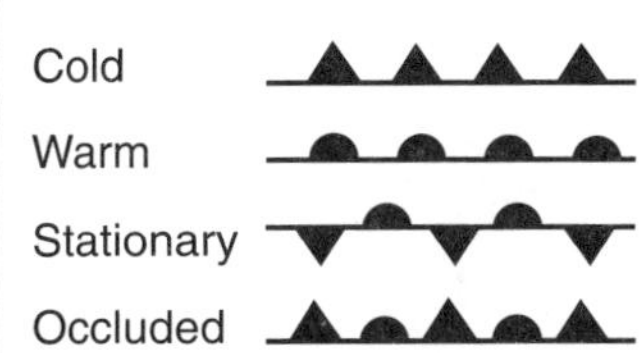

Hurricane

Tornado

Selected Properties of Earth's Atmosphere

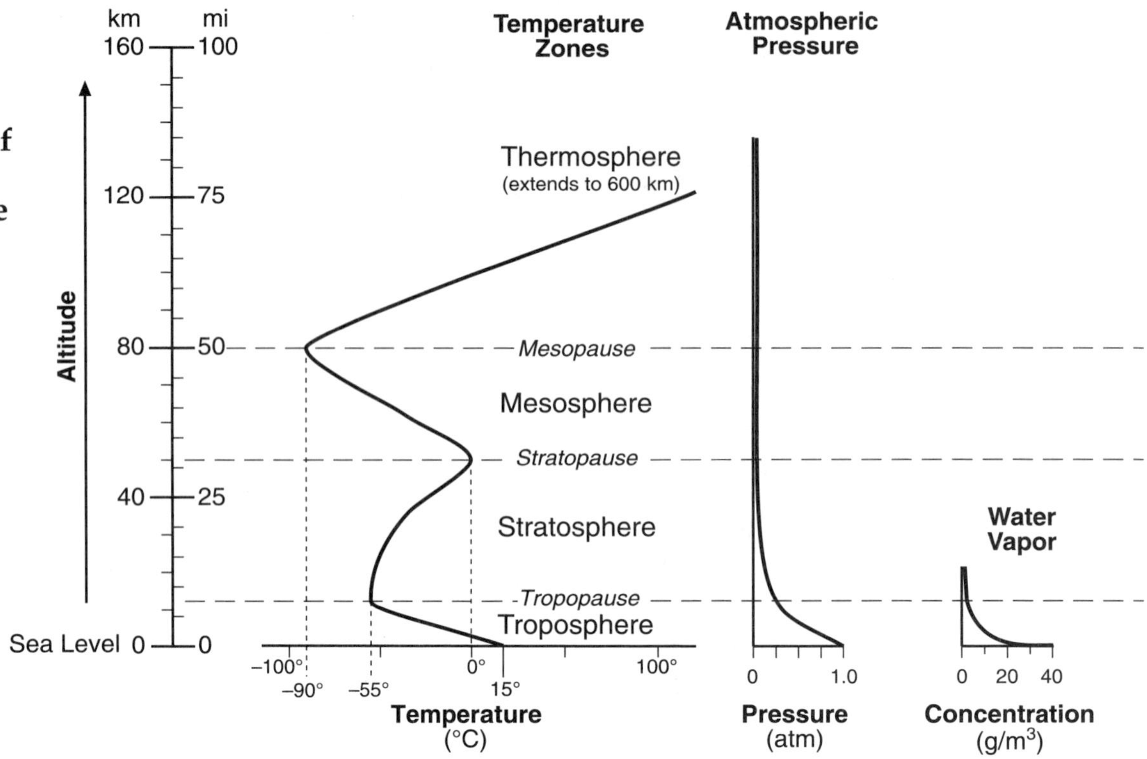

Planetary Wind and Moisture Belts in the Troposphere

The drawing on the right shows the locations of the belts near the time of an equinox. The locations shift somewhat with the changing latitude of the Sun's vertical ray. In the Northern Hemisphere, the belts shift northward in the summer and southward in the winter.

(Not drawn to scale)

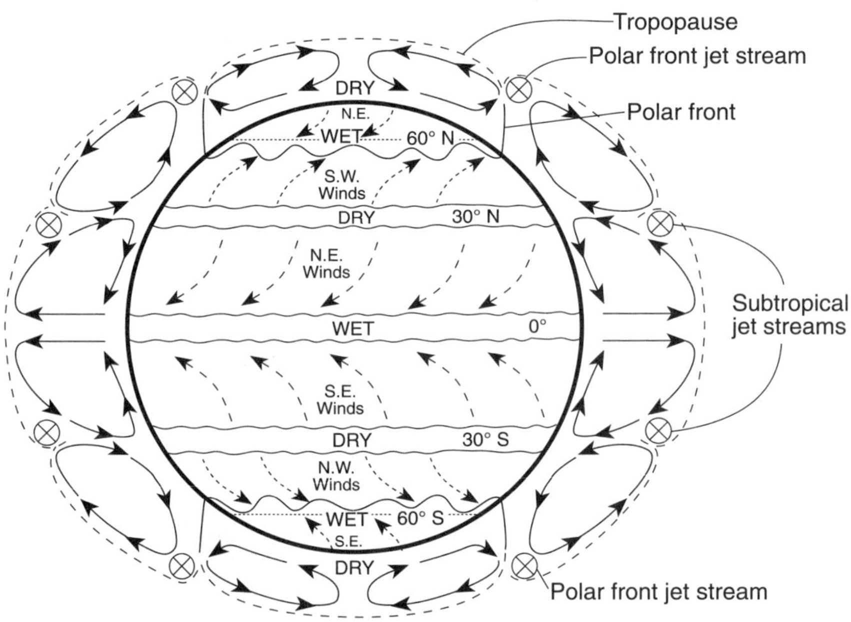

Electromagnetic Spectrum

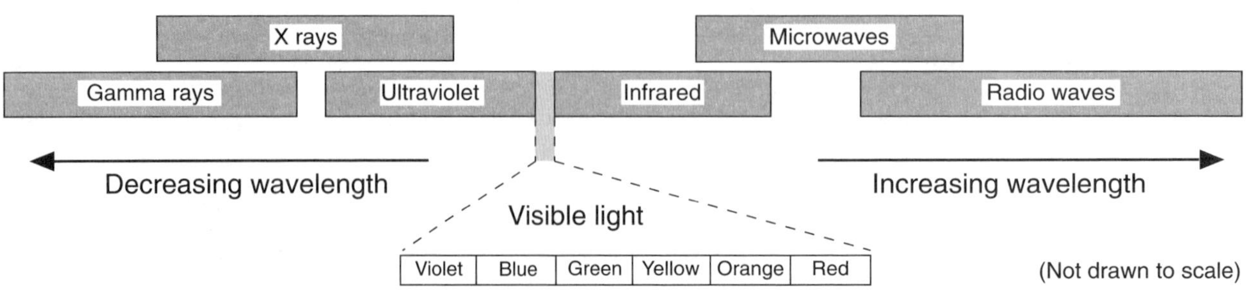

(Not drawn to scale)

Characteristics of Stars

(Name in italics refers to star represented by a ⊕.)
(Stages indicate the general sequence of star development.)

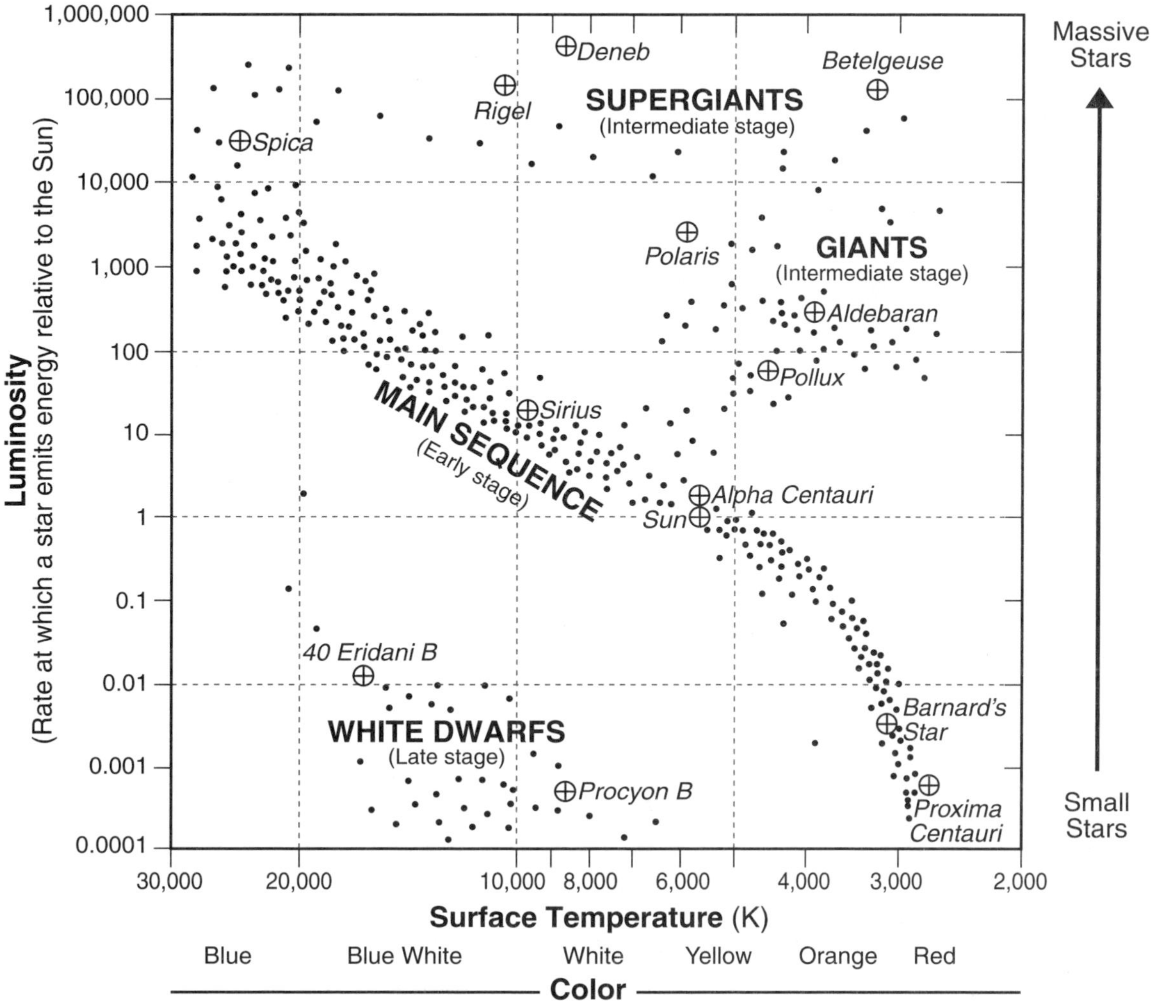

Solar System Data

Celestial Object	Mean Distance from Sun (million km)	Period of Revolution (d=days) (y=years)	Period of Rotation at Equator	Eccentricity of Orbit	Equatorial Diameter (km)	Mass (Earth = 1)	Density (g/cm^3)
SUN	—	—	27 d	—	1,392,000	333,000.00	1.4
MERCURY	57.9	88 d	59 d	0.206	4,879	0.06	5.4
VENUS	108.2	224.7 d	243 d	0.007	12,104	0.82	5.2
EARTH	149.6	365.26 d	23 h 56 min 4 s	0.017	12,756	1.00	5.5
MARS	227.9	687 d	24 h 37 min 23 s	0.093	6,794	0.11	3.9
JUPITER	778.4	11.9 y	9 h 50 min 30 s	0.048	142,984	317.83	1.3
SATURN	1,426.7	29.5 y	10 h 14 min	0.054	120,536	95.16	0.7
URANUS	2,871.0	84.0 y	17 h 14 min	0.047	51,118	14.54	1.3
NEPTUNE	4,498.3	164.8 y	16 h	0.009	49,528	17.15	1.8
EARTH'S MOON	149.6 (0.386 from Earth)	27.3 d	27.3 d	0.055	3,476	0.01	3.3

Properties of Common Minerals

LUSTER	HARD-NESS	CLEAVAGE	FRACTURE	COMMON COLORS	DISTINGUISHING CHARACTERISTICS	USE(S)	COMPOSITION*	MINERAL NAME
Metallic luster	1–2	✔		silver to gray	black streak, greasy feel	pencil lead, lubricants	C	**Graphite**
	2.5	✔		metallic silver	gray-black streak, cubic cleavage, density = 7.6 g/cm^3	ore of lead, batteries	PbS	**Galena**
	5.5–6.5		✔	black to silver	black streak, magnetic	ore of iron, steel	Fe_3O_4	**Magnetite**
	6.5		✔	brassy yellow	green-black streak, (fool's gold)	ore of sulfur	FeS_2	**Pyrite**
Either	5.5 – 6.5 or 1		✔	metallic silver or earthy red	red-brown streak	ore of iron, jewelry	Fe_2O_3	**Hematite**
Nonmetallic luster	1	✔		white to green	greasy feel	ceramics, paper	$Mg_3Si_4O_{10}(OH)_2$	**Talc**
	2		✔	yellow to amber	white-yellow streak	sulfuric acid	S	**Sulfur**
	2	✔		white to pink or gray	easily scratched by fingernail	plaster of paris, drywall	$CaSO_4 \cdot 2H_2O$	**Selenite gypsum**
	2–2.5	✔		colorless to yellow	flexible in thin sheets	paint, roofing	$KAl_3Si_3O_{10}(OH)_2$	**Muscovite mica**
	2.5	✔		colorless to white	cubic cleavage, salty taste	food additive, melts ice	$NaCl$	**Halite**
	2.5–3	✔		black to dark brown	flexible in thin sheets	construction materials	$K(Mg,Fe)_3AlSi_3O_{10}(OH)_2$	**Biotite mica**
	3	✔		colorless or variable	bubbles with acid, rhombohedral cleavage	cement, lime	$CaCO_3$	**Calcite**
	3.5	✔		colorless or variable	bubbles with acid when powdered	building stones	$CaMg(CO_3)_2$	**Dolomite**
	4	✔		colorless or variable	cleaves in 4 directions	hydrofluoric acid	CaF_2	**Fluorite**
	5–6	✔		black to dark green	cleaves in 2 directions at 90°	mineral collections, jewelry	$(Ca,Na)(Mg,Fe,Al)(Si,Al)_2O_6$	**Pyroxene** (commonly augite)
	5.5	✔		black to dark green	cleaves at 56° and 124°	mineral collections, jewelry	$CaNa(Mg,Fe)_4(Al,Fe,Ti)_3Si_6O_{22}(O,OH)_2$	**Amphibole** (commonly hornblende)
	6	✔		white to pink	cleaves in 2 directions at 90°	ceramics, glass	$KAlSi_3O_8$	**Potassium feldspar** (commonly orthoclase)
	6	✔		white to gray	cleaves in 2 directions, striations visible	ceramics, glass	$(Na,Ca)AlSi_3O_8$	**Plagioclase feldspar**
	6.5		✔	green to gray or brown	commonly light green and granular	furnace bricks, jewelry	$(Fe,Mg)_2SiO_4$	**Olivine**
	7		✔	colorless or variable	glassy luster, may form hexagonal crystals	glass, jewelry, electronics	SiO_2	**Quartz**
	6.5–7.5		✔	dark red to green	often seen as red glassy grains in NYS metamorphic rocks	jewelry (NYS gem), abrasives	$Fe_3Al_2Si_3O_{12}$	**Garnet**

*Chemical symbols:

Al = aluminum	Cl = chlorine	H = hydrogen	Na = sodium	S = sulfur
C = carbon	F = fluorine	K = potassium	O = oxygen	Si = silicon
Ca = calcium	Fe = iron	Mg = magnesium	Pb = lead	Ti = titanium

✔ = dominant form of breakage